PLANET HEART

FRANÇOIS REEVES, MD
Translated by JOAN IRVING

planet HEART

HOW AN **UNHEALTHY ENVIRONMENT** LEADS TO **HEART DISEASE**

David Suzuki Foundation
GREYSTONE BOOKS
Vancouver/Berkeley

Translated from the original French, *Planète Cœur*, by Dr. François Reeves

14 15 16 17 18 5 4 3 2 1

Greystone Books Ltd.
www.greystonebooks.com

David Suzuki Foundation
219–2211 West 4th Avenue
Vancouver BC Canada V6K 4S2

Cataloguing data available from Library and Archives Canada
ISBN 978-1-77100-081-9 (pbk.)
ISBN 978-1-77100-082-6 (ebook)

Editing by Tom Holzinger and Iva Cheung
Copy editing by Iva Cheung
Cover and text design by Jessica Sullivan
Cover photo by iStockphoto.com
Interior image credits on page 176
Printed and bound in Canada by Friesens
Distributed in the U.S. by Publishers Group West

We gratefully acknowledge the financial support of the Canada Council for the Arts, the British Columbia Arts Council, the Province of British Columbia through the Book Publishing Tax Credit, and the Government of Canada through the Canada Book Fund for our publishing activities.

Greystone Books is committed to reducing the consumption of old-growth forests in the books it publishes. This book is one step toward that goal.

contents

one

SURGICAL SHARPSHOOTER

THE ELITE MARKSMAN and the interventional cardiologist each has the same goal: to lodge a metal cylinder into the human heart.

For one, it's a bullet; the other, a stent. One explodes the heart; the other makes it whole again.

They stand as symbols of the contradictory potential of human nature. These two experts go through similar rigorous, lengthy, and near-obsessive training. They spend thousands of hours repeating and tirelessly fine-tuning specific movements, to make the eye sharper, the hand surer. They test every advance in technology and knowledge, to expand skills acquired during years of hands-on work, years of peering at the infinitely distant and the infinitesimally small.

Ultimately, when these two experts fire their bullets at the heart, they almost always find their mark, but the end results couldn't be more different: one, death; the other, life.

This image is what sometimes comes to mind when I am summoned from bed at two in the morning to treat a heart attack: my colleagues and I in interventional cardiology are New Age sharpshooters. A whole SWAT team is called into action—paramedics, technicians, nurses, emergency teams, cardiologists—all putting their skills to work to implant a metal alloy tube into a blocked artery and thereby save a life. Our mission is to beat the clock, defy death and time, and conquer the disease.

We begin furtively at first, with a local anesthetic and a sedative. We introduce our catheter through the wrist artery and thread it to the coronary arteries of the heart, wherein the problem lies. Within minutes we clear the blocked artery and secure our stent there to keep it open. Meanwhile we continue to speak quietly to the victim of the attack. Released from the excruciating chest pain, the patient gradually becomes calmer from the sedation and the repair of the damaged artery.

It is a scene we find ourselves in at all hours of the day and night, involving all kinds of unanticipated encounters—with cabinet ministers and comics, police officers and drug addicts. The patient speeds along streets under the flashing red light of the ambulance, the standby team is roused from their dinner tables or beds, and all rush toward the cardiac catheterization laboratory. A team of twenty is about to work as one.

And what is the cause of all this commotion? An atherosclerotic plaque has chosen this moment to break away from the inner wall of a vital artery, unleashing a tsunami into the lives of both victim and healers.

A Worldwide Tsunami of Disease

There are seventy-two coronary sharpshooters in Quebec, the province where I practice medicine. We interventional cardiologists work in fourteen cardiac catheterization centers, each open twenty-four hours a day, seven days a week. We operate in teams alongside cardiovascular nurses and X-ray technicians and are supported by the staff of emergency rooms and coronary care units. In 2009 our teams treated 5,000 acute myocardial infarctions—heart attacks—and we performed 38,000 catheterizations, including 16,000 angioplasties with stent placement. In addition, our cardiac surgeons performed 6,500 heart bypass procedures when more than an angioplasty was needed to repair the damage.

In 2008, approximately 6.5 million diagnostic and therapeutic interventional cardiology procedures were performed in the U.S. That number was expected to climb to approximately 8.1 million in 2013, with corresponding sales of interventional cardiology products projected to reach a total value of nearly $6.7 billion.[1]

Cardiovascular disease (CVD) is the primary cause of death the world over. Each year more people die from CVD than from any other cause. In 2008 the World Health Organization (WHO) estimated that CVD is responsible for 17.3 million deaths annually or 30 percent of total deaths internationally. Among those deaths, an estimated 7.3 million are caused by heart disease and 6.2 million by stroke.[2]

The WHO also estimated that, by the year 2030, some 23.6 million people will die annually from some type of cardiovascular disease, mainly heart disease and stroke. According to WHO projections, these two diseases will remain the primary causes of death.

In developing countries, the incidence of such disease is exploding. There has been a four- and in some countries a sevenfold increase in cardiovascular disease in just two generations.[3]

Fingering the Culprits

Where does it come from, the grime in people's biological plumbing that threatens to stop the heart? Why do highly trained teams of medical specialists so often have to rush to save the life of someone who felt perfectly well just an hour before?

When I studied medicine, I was taught that there are five classic risk factors for CVD: heredity, smoking, high blood pressure, high cholesterol, and diabetes. Most of these are "silent killers"; their damage is not visible until years later. So how were they identified? By quiet, patient investigative work on a grand scale—the Framingham Heart Study—which makes a great detective story in its own right.

In 1948, the Framingham Heart Study embarked on an ambitious project in health research. At the time, little was known about the general causes of heart disease and stroke, but the death rates for CVD had been increasing steadily since the beginning of the century and had become an American epidemic.[4]

An American Epidemic

Heart attacks were relatively infrequent in the nineteenth century, but in the years following the Second World War, one-third of Americans over the age of fifty suffered a heart attack. The death rate from CVD was double that from cancer. The epidemic of heart disease was so great that life expectancy in the United States remained stuck at age forty-five, despite the numerous medical advances of the period, particularly the reduction in child mortality, discovery of antibiotics, and improvements in surgical procedures. The situation in Canada was much the same, according to figures published by Statistics Canada. No data are available for the years before 1920, but the death rate from cardiovascular disease rose sharply over the next three decades, from 221.9 deaths per 100,000 people in the early 1920s to a peak of 414 deaths per 100,000 people in the 1950s.[5]

The Framingham Heart Study

Alarmed, but confident in the strength of their organizational abilities and financial resources in the post-WWII era, American public health authorities decided to examine one city to discover why, over the age of fifty, one in three Americans had a heart attack. The city they chose was Framingham, a small urban center close to Boston, itself a hub for medical education and research whose universities were overflowing with many of the resources needed to carry out such a mega-study.

Investigators descended upon Framingham and drew up lists of families. Volunteer participants were invited in a systematic way—not randomly. In the end the official "Framingham cohort" was a group of 5,209 men and women aged thirty to sixty-two who were regularly interviewed and tested for CVD-related factors for more than forty years—in many cases until death.

The Boston-based epidemiologists were able to determine with some precision who developed CVD and who did not. Because they knew so much about the health history of their cohort, they were soon able to make correlations between the different factors in a person's health history and the onset of heart disease.

This complex—and invaluable—method was first pioneered by the Framingham researchers and remains in wide use today. Indeed, successors of the original team are now following a fourth generation of Framingham citizens, who generously continue to participate in it for the good of humanity.[6]

Since its launch in 1948 the Framingham Heart Study has published more than 1,200 scientific articles. One of its early researchers and later its director, William B. Kannel, was responsible for one of the most important breakthroughs in the history of medicine: he originated the concept of the *risk factor.*

Kannel coined the term "risk factor" in 1961 in a landmark publication in the *Annals of Internal Medicine* and promoted the concept that CVD is multifactorial in origin—that is, a combination of factors may contribute to and increase the risk that a CV event will occur.[7]

The model Kannel developed was the foundation for a whole new understanding of heart disease that, in turn, led to a significant reduction in the death rate from CVD in the developed world, though mainly in North America. In 1976 Kannel received a Canada Gairdner International Award for "careful epidemiologic studies, revealing risk factors in cardiovascular disease which have important implications for the prevention of these disorders."[8]

From Risk Factors to Physiological Causes

Before Kannel's work, the practice of medicine was rooted in the concept of cause and effect. You become infected by a Koch's bacillus, and you come down with tuberculosis. A meningococcus invades your body, and you get meningitis. The more flexible concept of risk factor was revolutionary when it was introduced; today it is ubiquitous.

In simple terms, the concept of risk works like this: You are driving from Montreal to Toronto at 200 kilometers per hour (120 miles per hour), so there is no doubt you will arrive at your destination in good time. However, there is a high risk that you will have an accident. In contrast, if you travel at 110 kilometers per hour (70 miles per hour), there is still a chance that you will crash, but the risk of such an accident happening is much lower than it would be at the higher speed. And if one thousand drivers are heading to Toronto at 200 kilometers per hour, there will be many more accidents than in a group of one thousand drivers doing 110 kilometers per hour. The *relative risk* of the latter group is lower.

Speeding is a risk factor for car accidents. So too are faulty brakes, poor road visibility, and drunken driving. If a number of these risk factors are present at the same time and the driver continues to travel, an accident becomes almost inevitable. Cardiovascular disease is a bit like car accidents—enough risk factors over enough time, and a heart attack seems virtually guaranteed.

The evolution of mathematics and biostatistics has made it possible to determine with reasonable precision the impact of health risk factors and to discriminate between different outcomes when several factors occur together. In order to explain why a risk factor (the culprit) has a correlation with a disease, physiologists must first determine why it is a threat (how the crime was committed). They begin by searching for the cellular mechanisms that explain how a particular risk factor helps cause a certain disease and then gradually fill in the details that lead to both a better understanding of the disease and better treatments for it.

The Three Triads

Based on our current understanding, the acknowledged risk factors for heart disease may be roughly grouped into three triads:

- Who I am
- What I do
- Where I live

FIRST TRIAD: WHO I AM

The CHD triad: cholesterol, hypertension, diabetes

It is the genetic makeup, or DNA, of an individual that determines his or her predisposition to dyslipidemia (an abnormal level of cholesterol or triglycerides in the blood), hypertension (high blood pressure), and diabetes. In the past few years the science of genetics has exploded, most recently through the work of the Human Genome Project, launched in 1990 by the U.S. Department of Energy's Office of Science. As I write, the world's foremost geneticists and biochemists are filling in the details of a blueprint of the human being, sort of a spec sheet for people. You can now view an online map of the entire human genetic code—which is pretty much your own code too.[9]

The genome map is full of surprises. Human DNA is made up of 30,000 genes, considerably fewer than the 80,000 to 140,000 researchers expected to find. They are complemented by long sections of regulatory DNA that switch the genes on and off.

Each gene is a long sequence of the four-letter genetic alphabet ACGT, or adenine, cytosine, guanine, and thymine. And every one of the 100 trillion cells in every individual (except the red blood cells, which lack a nucleus but carry the precious hemoglobin protein that brings oxygen to the body's tissues—see Chapter 14) contains 3.2 billion ACGT nucleic acid bases, the building blocks of our genes. As astronomical as these figures are, 99.9 percent of the base sequences are the same in every human being.

Despite these common traits, there are great differences among individuals with respect to the first triad: one person may present early-onset hypertension, another diabetes, and yet another high cholesterol or triglycerides in the blood. The predisposition to develop one of these conditions is linked to its corresponding genes—slight DNA variations that are passed down through families.

SECOND TRIAD: WHAT I DO

The SOT triad: sedentary lifestyle, obesity, tobacco

Even though genes are buried deep in each cell and are unchangeable—except in cases of accidental mutation after exposure to radiation or other mutagens—the *expression* of these genes is strongly influenced by lifestyle and environment. The phenomenon is so significant that it has spawned a new field of scientific study: *epigenetics*.

For instance, the cancers caused by smoking develop largely because of the substances (aptly called carcinogens) that disrupt genes and alter their expression. Scientists have also discovered that a variety of cancers result from dietary deficiencies and industrially modified foods and that a sedentary lifestyle further contributes to the disease. In addition—and this is an interesting parallel—many cancers have a set of risk factors that overlaps with those of heart disease.

Genes predispose but rarely act as a sole cause. The expression of any gene also needs just the right environment: genetically similar populations may have quite different rates of hypertension, diabetes, and high cholesterol depending on the specific environment in which they live.

Among thousands of examples, one of the most striking is that of the descendants of the residents of Okinawa. This prefecture in Japan is noted for the exceptional longevity of its population and the high number of centenarians, attributable in large part to excellent cardiovascular health.[10] Descendants of this group who immigrated to Hawaii, however, have twice the incidence of cardiovascular disorders, whereas those who immigrated to Los Angeles suffer a fourfold higher rate.

The subject was notably discussed on a website for citizen journalists in January 2007:

> Comment: I read that on one of the Pacific islands where residents have the same longevity, public health stats have already tanked because the American base is so near and the type of food eaten there has spread. But I don't remember where it is. The same phenomenon occurred among the Navajo Indians and people living in Canada's Far North.

> Reply: I agree entirely. That's what happened in Okinawa, the largest American base in the Pacific. After the defeat of the Japanese at the end of the Second World War, the American administration introduced its food preferences to schools. Today, the farther down on the age pyramid you are, the lower your life expectancy. The teenagers of Okinawa have the highest rate of obesity in all of Japan. The same thing is seen in Greece, Italy, and Spain.[11]

Such situations are precisely what led to the founding of epigenetics, which aims to help us understand how human genes are expressed as an outward response to the environment. More specifically, epigenetics helps the cardiologist to better understand why the rates of hypertension, diabetes, and high blood cholesterol vary according to where we live.

THIRD TRIAD: WHERE I LIVE

The EDU triad: environment, diet, urbanism

Statistics published by organizations such as the WHO and the Centers for Disease Control and Prevention (CDC) reveal startling differences in death rates from cardiovascular disease based on location. The nations of the former Soviet bloc have CVD death rates that are up to ten times the rates in Western Europe.[12] In Russia, the CVD death rate among men aged twenty-five to sixty-four is 762 per 100,000 inhabitants, in Ukraine it is 600, whereas in France and Norway it is roughly 70. Brutal! This difference of 1,000 percent can't be blamed on genes and smoking alone.

More findings: among the so-called developed countries, the risk of heart attack in an adult who was adopted as a baby varied according to where the adoptive family lived: in Japan, France, Canada, or the USA.[13] These differences confirm the importance of environment as a risk factor in cardiovascular disorders, as was also observed among the Okinawans who grew up in two different regions of the United States.

As noted earlier, also surprising are the changes in the incidence of cardiovascular disease in North America over time. Although relatively rare at the beginning of the twentieth century, CVD rose sharply thereafter, cresting in the 1950s before declining until the 2000s, when it again began to climb.

It is now clear from converging lines of evidence that environmental factors play a great role in the onset of disease. Sometimes—as is the case with diet—a randomized clinical trial can be used to prove that a suspect food is innocent or guilty. In most cases of environmental exposure, however, a controlled trial is impossible, unethical, or both. We can't expose children to cigarette smoke to prove that it later causes cancer!

Or take another important example of immediate concern: air pollution. Epidemiologists have shown that air pollution alone kills up to 20,000 Canadians each year, mostly from lung and cardiovascular disease. Recently we have discovered that air pollution also contributes to high blood pressure and metabolic syndrome. And air pollution is just one of the many pernicious consequences of industrialization.

Industrial foods are another. As we shall see in later chapters, we now have increasing evidence to link the types of food people eat with the prevalence of obesity, metabolic syndrome, and—in particular—diabetes, which are all significant factors in heart disease.

We also see a link between urbanism and human health. You have probably read articles or seen news reports that evaluate and describe cities in terms of their obesity neighborhoods, heat islands, heavily paved-over and polluted sites, and, on a more positive note, their proactive and heart-friendly environments.

A polluted urban environment, we now know, is associated with more deaths from cardiovascular disease than a green environment. Urban pollution also strongly exacerbates the *social* problems that contribute to heart disease: stress and negative interactions with the surrounding milieu.

The Three Triads: The Great Challenge

Back in the catheterization laboratory, the SWAT team has completed its work. The angioplasty is a success, the stent is in place in the repaired artery, the heart attack is stemmed, and the victim no longer faces death. The team relaxes and starts chatting as the patient is returned to the stretcher and wheeled to his room to be reunited with loved ones and family, to partake again of simple pleasures and passions. From the body's point of view another atherosclerotic plaque that has "gone rogue" has been successfully tamed.

And yet it is clear that we need to understand the reasons why this plaque appeared at all. Is it normal that the wall of a coronary artery should fill with atherosclerosis and become inflamed? Or perhaps such a plaque is the artery's response to a hostile, abnormal environment?

We can go further. Before industrialization, how prevalent was heart disease? Did early humans also have heart attacks? If we could transport our radiology and hemodynamics labs back to the Age of Antiquity, if we could do a heart catheterization of Tutankhamun, what would we discover?

two

PARADISE LOST

HUMAN BEINGS SEEM to have sacrificed the health of their arteries in the name of progress. But why should we worry? Life expectancy has never been so high. In France, it went from thirty-three years of age in 1800 to forty-eight in 1900 and then rose to seventy-nine years by the year 2000.

The significant increase in life expectancy is the result of improved public health services and also of diplomacy. Fewer epidemics and fewer wars. Louis Pasteur and others of his generation found ways to purify water and food. And in response to the horrors of war, world leaders pushed for mechanisms to ensure peace—first the Geneva Convention and later the United Nations.

After the Second World War, the focus again shifted to civilian life. In the United States, the U.S. Public Health Service discovered that deaths from cardiovascular disease had tripled between 1900 and 1950. It began designing a response, one aspect of which was the previously mentioned Framingham Heart Study. Launched with millions of dollars in financial support, this study attracted some of the best scientists then working in Boston.

In hundreds of reports, the Framingham study listed the reasons people develop heart disease. Since 1970 it has been widely accepted that the risk factors for atherosclerosis are heredity, high blood pressure, diabetes, high cholesterol, and tobacco use. In the 1990s two more factors were added to the list: obesity and a sedentary lifestyle.

Framingham, with its 1,200 published articles to date, has become the world's most important source of information on cardiovascular health. It is nonetheless fascinating to return to the very first article about the project—the one that launched the great endeavor—in particular, to the following excerpt:

> The word "epidemiology" refers to the study of something "which is thrust upon the people." Most workers agree that epidemiology deals with the fundamental questions as to where a given disease is found, where it thrives, where and when it is not found... in other words it is the ecology of disease.[1]

The ecology of disease. So the seeds of environmental cardiology were sown in a visionary article published in 1951, though perhaps the authors should have worded it "ecology and disease."

Ford Model T and Coca-Cola

As important as it is, the Framingham Heart Study has a major weakness and thus seems to have failed in its primary goal: to explain the devastating rise in heart disease in the United States in the first half of the twentieth century—a period during which the genetic makeup of Americans did not change in any significant way.

The end of the nineteenth century and early twentieth century marked the peak of the industrial revolution in North America, the era that saw the invention of the Ford Model T and Coca-Cola, symbols of the American Dream. Human life was transformed by industrialization, particularly by the increased dependence on fossil fuels and a dietary shift to factory-produced foods.

It wasn't until the twenty-first century arrived that doctors began to realize the extent to which the American Dream represented poison to human arteries. In hindsight, the weakness of the Framingham study was evident. *The residents of Framingham all lived in the same environment.* It was impossible to isolate the influence of environmental factors on participants' health (though these factors were not part of the survey in any case).

So before the industrial era—what was the state of our arteries?

The Secret of the Mummy and Tutankhamun's Heart

Egyptian civilization has long fascinated us for its extraordinary monuments, its sophisticated cult of the dead, and its three-thousand-year history. The Great Pyramid of Giza is the only one of the seven wonders of the ancient world that still stands, and mummies are among the oldest and rarest of human specimens on Earth. They are also reasonably well preserved. Thanks to modern imaging technologies, it's possible to study mummies with a minimum of tampering of their bodies.

In 2009 a group of anthropologists and radiologists performed whole-body computed tomography (CT) scans on twenty-two mummies from the Egyptian Museum in Cairo.[2] The "patients" had lived between the nineteenth and third centuries BC. Hardening of the arteries was found in 31 percent of the subjects and in 87 percent of those who died after the age of forty-five. The researchers concluded that atherosclerosis existed long before the industrial era. They also theorized that the subjects' "atherosclerosis" resulted from their privileged lifestyle. As members of the elite, these Egyptians had undoubtedly consumed large quantities of red meat.

Several commentators, however, called for more caution regarding the conclusions of the study. They pointed out that hardening of the arteries does not necessarily indicate the presence of atherosclerosis—other diseases can lead to similar deposits of calcium on arterial walls—and that the researchers had been unable to directly examine the mummies' arteries.

Nonetheless, any need for greater care in interpreting the subtleties of the results doesn't make the important information in the study any less valid. Close examination of the CT images reveals that there is very little calcification in the mummy arteries when measured against what my colleagues and I regularly encounter in today's cardiology patients. In fact, the small hardened plaques seem almost negligible compared with what we treat every day.

Tsimane and the Original Heart

Other people in other areas also have much to reveal about the human heart, especially those who live beyond the reach of industrial civilization. Scientists have suggested that these peoples have much in common with original humans and that they represent an opportunity to study the human body unaffected by industrialism. One group of great interest to pathologists, for example, are aboriginal Australians. But the religious beliefs and burial customs of aboriginal Australians prohibit autopsies and so don't allow us to examine the blood vessels of the dead.

In contrast, much information has been gathered from observing the Tsimane people of Bolivia. The Tsimane live along the banks of the Amazon and still practice a subsistence lifestyle largely uninfluenced by modern technology and consumerism. Michael Gurven, an anthropologist from the University of California Santa Barbara, has spent several years studying and writing about the Tsimane. The findings of Gurven and his team are extremely interesting.[3]

There is practically no peripheral arterial disease (PAD) among the Tsimane people (as measured by the difference in systolic blood pressure at the ankle and elbow—the ankle-brachial index) and very little hypertension. The rates of arterial disease and hypertension are not only lower in this group than in peoples from other countries (including the United States, South Africa, Spain, Sweden, and China) surveyed in the study; they are almost non-existent.

Surprisingly, arterial disease remains low even among elderly Tsimane. The hypertension rate of 10 to 20 percent for Tsimane elders in their seventies is similar to or lower than that of people in their forties in other countries.

Gurven and his colleagues proposed that this low rate of arterial disease is linked to the balanced metabolism, active lifestyle, healthy body mass, and low-fat diet of the Tsimane. Their diet includes a lot of fish as well as fruits and vegetables. It is also low in saturated fats and rich in potassium.

In addition the Tsimane are very active physically. This trait, according to the researchers, is a protective factor. Despite chronic inflammation from frequent, untreated infections, the Tsimane people have healthy arteries. Their arterial systems are protected by the active lives they lead and the food they eat.

These findings have allowed anthropologists to draw a startling conclusion. In the words of Michael Gurven:

> The absence of PAD and CVD among Tsimane parallels anecdotal reports from other small-scale subsistence populations and suggests that chronic vascular disease had little impact on adult mortality throughout most of human evolutionary history.[4]

There is always a *but*. In the case of the Gurven study, no data were collected on key environmental variables in the Tsimane region. The Tsimane people live deep in the Amazonian rain forest—sheltered from exhaust fumes and smog and beyond the reach of glucose-fructose syrup and trans fats.

Is it a paradise lost?

three

HUMANS ARE HOST PLANETS

OUR KNOWLEDGE OF the risk factors for cardiovascular disease is still evolving. From the era of the Framingham Heart Study, which identified the classic risk factors, we have passed into the era of David Suzuki, which calls for people everywhere to know and protect the environment.[1] As we have seen, the environment plays a large role in determining our susceptibility to heart disease. It influences both the ways our genes are expressed and our behavior, and it is a crucial factor in how and why our internal plumbing becomes clogged.

Before the twentieth century, humans' greatest adversaries were biological. Bears and tigers were common predators, as were other human beings, especially during wars and genocides. There were also micro-adversaries: microbes such as bacteria and viruses. Then industrialization unleashed new predators—the chemical and molecular ones. Today, it's these nano-adversaries that we must fear, the majority of which humans have introduced into nature themselves.

Ourselves as Planets

For the past three centuries, telescopes have vainly probed the skies and space in search of extraterrestrials—life beyond this planet. At the same time, however, the microscopes of Louis Pasteur, Edward Jenner, Alexander Fleming, and Robert Koch, among others, were revealing colonies of hitherto invisible aliens living not only *around* us but also *on* us and *inside* us. Scientists began to observe and understand that these microscopic predators cause numerous diseases; that they make people sick with pneumonia, bubonic plague, and typhoid; and that they could wipe out entire towns and cities.

Think about it this way: human beings are celestial bodies colonized by populations of up to 100 trillion bacteria. These bacteria inhabit every external and internal surface of the body. They swarm over the skin, which is approximately 1.8 square meters (19 square feet) in area, or about the same size as a tabletop. They also grow profusely in the digestive tract, which is 400 square meters (4,300 square feet), the equivalent of two tennis courts. Indeed, the survival of our "host planet" depends on the balance it strikes with its microbial residents.

Although they are the source of many diseases, bacteria are also essential to us. Some are even quite beneficial, as proven by the bacterial products wine and cheese, and other bacteria make a number of valuable medications. Healthy bacteria on the body block dangerous invaders, just as a flower bed full of vibrant perennials can ward off unwanted plants.

Bacteria transform the intestine into a perfect composter. They add their action to the chemical agents in our digestive juices. The relationship between such bacteria and humans is symbiotic. In an intestine or in a composter, the resident bacteria break down complex substances into small components that are used in other life-sustaining processes.

The human body is not unique in nature for having a vital mutual dependency with microorganisms. Trees do as well. In the forest, scores of mushrooms are attached to the subterranean root systems of plants, including the most massive trees. These mushrooms decompose the humus in which they grow and transform it into substances that are useful to their plant partners.

The unseen microbial residents of the human body synthesize such nutrients as vitamin B_{12}. Bacteria make bodily wastes biodegradable. They could well serve as a model for the global management of domestic and industrial waste. On a smaller scale, they make up a beneficial and perennial flor a that thrives in our interior "flower beds" to keep out undesirable parasites. Public sanitation and health measures have not eradicated bacteria—far from it. But by controlling the number and type of bacteria in circulation, they have made bacteria "socially acceptable," to the extent that, in some ways, bacteria now serve to prevent epidemics. Such measures enhance the wonderful contribution of bacteria and fungi to our lives.

Public health programs promoting disease prevention, hygiene, and vaccination have played a larger role in eradicating epidemics, and their horrific consequences, than even antibiotics. There are no antibiotics or other treatments for smallpox and polio. But these are examples of diseases under control.

Micro-Adversaries and Unfinished Symphonies

Jenner, Pasteur, and similar researchers had a hard time convincing their fellow citizens that invisible organisms—the micro-adversaries also known as bacteria, viruses, and fungi—attack and kill with the same deadliness as rogue bears and bullets.

In cardiology, the principal infectious diseases are acute rheumatic fever and endocarditis. In the West, between 1950 and

2000, thousands upon thousands of heart valves were dilated, operated on, repaired, or replaced. The cause of their malfunction: a group A streptococcal bacterium more commonly known as *strep A*. This bacterium is particularly loathsome because it attacks small children and can infect them with a dangerous fever of the heart and joints called acute rheumatic fever (ARF). It is very moving to hear older patients speak about ARF attacks they lived through in the years before antibiotics were discovered. A case in point: a five-year-old child riddled with joint pain so severe that he was confined to his bed for periods of up to a year. This youngster was deprived of school, friends, and games; his constant companion was the pain throughout his body.

In addition to destroying the lives of thousands of children, strep A leaves behind a time bomb. It induces a stealthy, chronic inflammation whose consequences surface decades later. Slowly but inexorably, strep A attacks the heart valves, which become damaged, scarred, and hardened, and which eventually stop functioning, causing heart failure. In the past people simply died. One of the most widely held theories is that Mozart, who died at thirty-five, suffered from rheumatic fever.[2] Regrettably, the disease ended both his life and his prodigious artistic output. Had penicillin been available at the time, Mozart would likely have gone on to create hundreds more hours of sublime music.

Today, ARF is almost eradicated in the developed countries, virtually a relic of the past. In Canada, barely two hundred cases of ARF appear each year. Normally they are quickly diagnosed and treated and have no lasting complications—all because of the miracle of antibiotics. In the developed countries, cardiologists now rarely see and perform surgery on patients with valve disease caused by ARF, and it is only a matter of time before ARF completely disappears. In the emerging countries, however, the disease is still rampant. In China, India, and Russia, hundreds of thousands of people come down with ARF every year.

The same is true for bacterial endocarditis. Cases of this infection of the heart valve, which once led almost invariably to death, are now few and far between. Another sufferer of ARF in early childhood, Gustav Mahler, died of endocarditis at the age of fifty,[3] leaving behind his unfinished Tenth Symphony. His death deprived the world of many more years of his musical genius.

Viruses and *The Matrix*

The world of viruses is even stranger than the world of bacteria, and the aggressors are even smaller.

Viruses often seem bizarre. Not quite living beings themselves, they exist on the frontier of life. What's their purpose? Why do they exist? Probably simply because they can.

Viruses represent life in its most stripped-down form. They are fragments of DNA or RNA that move around in other organisms with a single goal: to find a vector by which to reproduce because they can't do so on their own. The virus must first infect another cell and then insert its DNA into the nucleus of this cell so that the host nucleus may copy it.

Biodiversity is characterized by equilibrium between environments and food chains, but viruses don't seem to have a place within this grand scheme. Viruses are, apparently, incapable of equilibrium. They can't enter into a treaty of coexistence like the one between humans and bacteria. For them, one must die. Existence is tied to a single issue: their death or the death of the host cell.

Such deadly parasitism is widespread in nature. Take the Common Cuckoo. In order to enhance the chances of survival of its offspring, the female lays eggs in the nests of other species (one egg per nest). When this chick hatches, it pushes the young of the host species out of its own nest!

A classic description of this pattern of behavior comes up in the film *The Matrix*, as explained by Agent Smith to Captain Morpheus:

> Every mammal on this planet instinctively develops a natural equilibrium with the surrounding environment, but you humans do not. You move to an area and you multiply and multiply until every natural resource is consumed. The only way you can survive is to spread to another area. There is another organism on the planet that follows the same pattern. Do you know what it is? A virus. Human beings are a disease, a cancer of this planet. You are a plague, and we are the cure.[4]

At one extreme is life in its most complex form, the human being; at the other is the most rudimentary form of life, the virus. They each exhibit the same environmental behavior.

Twenty-First-Century Nano-Aggressors

Building on the work of previous centuries, medical investigators of the twenty-first century continue their research to expand our knowledge of the threats to human health. In recent years specialists in environmental health have made us aware of aggressors even stealthier and smaller than bacteria and viruses. Although they are not alive, they are omnipresent in the environment. We know them as molecules and particulates, and they are shaping up to be a new line of assailants: the cardiovascular nano-aggressors.

Today's developments in chemistry build on the discoveries of seventeenth- and eighteenth-century scientists such as the Irish researcher Robert Boyle and the French chemist Antoine de Lavoisier. The technological breakthroughs of modern industrial chemistry have yielded more than 120,000 synthetic molecules that have, in the short term, proven very effective in improving our quality of life. The challenge is to understand how they will affect us over the long term.

It's time to take a look at our molecular cuisine.

four

MOLECULAR CUISINE

KNOWLEDGE ABOUT THE molecules we eat—molecular cuisine and molecular gastronomy—is a popular topic nowadays. The concept here is really to rediscover the age-old art of eating well. Enthusiasts want to know why some meals have the power to overwhelm with delight, so they cook "scientifically" to achieve a particular goal. Their pursuits stem from those most-enduring of human traits: curiosity and the desire to improve upon the knowledge of their predecessors.

In the nineteenth and twentieth centuries, chemists identified *vitamins*, essential molecules that the human body can't produce by itself and must therefore get from nature. Their observations and research were motivated by human needs and deficiencies, a permanent font of scientific endeavor. Much time and energy were devoted to finding the causes of the deadly diseases of beriberi (vitamin B_1 deficiency) and scurvy (vitamin C deficiency). The latter afflicted almost all of the men of Jacques Cartier's crew during their terrible first winter in Canada. About one-third of them died, not from freezing temperatures but from scurvy.

Thanks to the science of human physiology, the vitamins and trace minerals essential to our bodily health are now well known: iron, magnesium, copper, and so on. The breakthroughs in this field are celebrated in French author Rémi Cadet's *L'invention de la physiologie* (The invention of physiology), a marvellous book that should be required reading in every faculty of medicine. Cadet examines key moments in the framing of medical science's great theories, bringing this vast subject alive with many entertaining anecdotes about long-held beliefs and medical ideas spanning three centuries.[1] So many dogmas dethroned!

Since then, many other properties of food have come to light. We are discovering, for example, that certain foods have antioxidant and even anticarcinogenic (anticancer) benefits. We are taking a growing interest in the lycopene in tomatoes, the omega-3 fatty acids in flaxseed, and the antioxidant properties of blueberries. Despite many centuries of accumulated knowledge in biology, we are still doing research on herbal products, and we are only now really unraveling the facts about biodiversity and the health-promoting qualities of certain proteins.

We will, of course, meet with obstacles and detours as we go down the path of seeking better nutrition for human beings. We will encounter many who, without any scientific proof, make questionable or even laughable claims about the health benefits of certain food ingredients. Homeopathy and numerous so-called natural products have been mired in such controversies.

It's too bad that there has been so little definitive research in this field because scientists need objective knowledge, doctors need appropriate drugs, and consumers need accurate information to help guard themselves against the shameless, relentless charlatans of miracle foods. Magical thinking is tenacious!

More and more cookbooks and diet books are published every year and receive a barrage of media attention. Because there is no need for yet another one, the next few chapters will not cover the details of our molecular cuisine that are already well studied and publicized. Instead we will examine another important factor: the molecules we should *not* eat. What we really need to know in choosing our food is which foods to avoid.

In surveying the world and its infinite variety of foods, no particular cuisine jumps out as better and healthier than another. Asian, African, and European cuisines are all good if eaten according to certain basic principles—namely, that you eat a variety of high-quality foods and you eat in moderation. Millet, for instance, is the equivalent of bread, which is the equivalent of pasta. The real problem isn't the food itself but the industrial processes used to transform it.

We know that the diet of Canada's famed lumberjacks of yesteryear contained huge amounts of high-calorie fat. Today's nutritionists would be horrified at what those men ate. Yet lumberjacks had amazing physical characteristics and phenomenal strength compared with, say, their sedentary grandsons. Lumberjacks who ate 5,000 calories at a single meal—a meal for an Olympic athlete—burned all those calories in a day's work. They rarely suffered from high blood pressure, high cholesterol, diabetes, or cardiovascular disease. Because they worked at high intensity and in harsh climatic conditions, they were able to metabolize very high caloric loads. Anyone eating the same diet today would become morbidly obese.

Our lumberjacks never had the opportunity to eat turmeric or drink green tea, products imported from faraway places at great cost. Then again, they were never subjected to air pollution, ozone, particulate matter, cola drinks, high-fructose corn syrup, salt-laced vegetable juices, fried foods and pastries heavy with trans fats, hormone-treated beef, tartrazine, aspartame, and phosphoric acid. Despite the copious amounts they ate, few lumberjacks contracted cardiovascular disease—and not only because of their strenuous physical activity.

The Fruits of Industry

Following those early years, an industrialized food supply was developed to combat and eliminate an age-old scourge: lack of access to food. The solution to the problem seemed to lie in improved food preservation techniques and more efficient food handling and shipping. The solution *seemed* to be linked to the development of food additives that would make such advances possible.

But soon, different additives were developed to alter foods in other "useful" ways. Some made food tastier, others made it smoother, and yet others made it more presentable. Additives became a weapon of the food marketers, enabling manufacturers to pull out all the stops in refining the look and taste of their foods and differentiating their products from those of competitors. Soon consumers had no idea what was in the packages of food they bought.

In recent years this situation has improved, owing to food labeling regulations that require all ingredients in a product to be listed. A good thing, because now we have at least a partial answer to the question, What is it that we are eating?

Although there are hundreds of harmless food additives, other additives corrupt the wholesome bounty of the land when slipped in by the food industry. I call those additives "malware molecules." Widely used in restaurant fast foods as well as in mass-produced foods, they help cause metabolic syndrome (characterized by overweight, high blood pressure, diabetes, high cholesterol, and high triglycerides) and are directly linked to cardiovascular disease.

Russia, India, and China have all experienced a recent surge in cardiovascular disease, despite the fact that, historically, strokes and heart attacks were relatively rare there. In just two generations the burden of cardiovascular disease in these three countries has risen to equal or exceed that of the West. This explosion of cardio-metabolic syndrome and cardiovascular morbidity are simply collateral damage from the rapid worldwide spread of the Western industrial model. The two biggest culprits in this scenario are in what we breathe and what we eat: the malware molecules found in air pollution and fast food.

Politics and Diet

In 2010, the European Union (EU) considered an amendment to the law governing food labeling in EU member states. Some members were pushing for what is called a "traffic light" system, in which packaged foods sold in stores would carry color-coded labels advising consumers of the food's health benefits or risks based on its proportions of salt, sugar, and trans and saturated fats. Under the proposed system, a green sticker would mean a food that was always healthy; a yellow sticker, one to be eaten in moderation; and a red sticker, one to be avoided. The vote on the legislation was defeated by just two votes, thirty-two to thirty.

I'd like to make the choices even starker. Here's a list of what I call the malware molecules—the ones we need to avoid:

- high-fructose corn syrup
- phosphoric acid
- trans fats
- excess salt
- saturated animal fats

And here are the somewhat less harmful molecules—the ones to be consumed with caution:

- aspartame
- food coloring

- simple sugars
- alcohol (for women, more than two drinks per day or more than ten per week; for men, more than three drinks per day or more than fifteen per week)
- genetically modified foods (GM foods)

Let's go on a cook's tour of some of these items.

five

"OPEN HAPPINESS"

THERE IS A TOUCHING scene in the blockbuster film *Slumdog Millionaire* in which a young man opens a bottle of Coca-Cola and hands it to a pair of pre-adolescent street kids for a sip to clear their hot and dusty throats. To children as miserable and malnourished as these two, the gift of a soft drink is nothing short of a miracle. But the gesture has nothing to do with compassion: the young hustler plans to win the children's love and trust, then mutilate them horribly and force them to beg for him in the streets of Mumbai.

Although we live far from the poverty of the developing world, soft drinks play a nefarious role in our own society, as we shall see.

Since 1928 the Coca-Cola Company has been both a sponsor and an exclusive supplier to the Olympic Games since 1928: its contract as the supplier of non-alcoholic beverages to the Olympics runs until 2020. During the Vancouver Winter Olympics in

2010, the company inundated the airwaves with its "Open the Games, Open Happiness" advertising campaign, designed to persuade consumers that the values of the Coca-Cola Company mesh perfectly with those of the Olympics and that happiness and optimism are paramount in life.

Happiness and optimism? Certainly not among cardiologists. But let's not make the debate personal or stigmatize one or another soft-drink manufacturer. Rather, let's examine the scientific literature and consider the health impacts of soft-drink consumption.

As far back as 1942 the American Medical Association recommended that consumers limit the amount of sugar in their diets, in particular the number of soft drinks they consume. Its recommendation did not have much impact. In 1942 the average American drank ninety servings (240 milliliters or 8 ounces per serving) of soda annually. Fast-forward to the year 2000, when consumers drank an average of six hundred servings of this size each year.[1]

Reports confirming the potentially harmful effects of soft drinks began appearing in the 1980s. A survey of 59,000 African-Americans showed that drinking two soft drinks per day increased the incidence of diabetes by 24 percent.[2] Another study, entitled the Nurses' Health Study II, followed 51,603 women for eight years and found that participants who drank one or more soft drinks daily were 83 percent more likely to gain 1 or more kilograms (2.2 or more pounds) of weight per year and to develop diabetes.[3] A study of pregnant women found that those who drank more than five soft drinks a week were 22 percent more likely to have diabetes than those who drank one or fewer soft drinks a month. A diagnosis of gestational diabetes means that a pregnancy is at high risk and that both mother and child are at risk of complications.[4]

A 2007 review study by Yale University analyzed the results of ninety-eight studies on sugar-sweetened beverages, nutrition, and health.[5] The findings were eye-opening:

- Between 1970 and 1997 the annual consumption of soft drinks per capita in the United States rose 87 percent, from 83 liters (22 gallons) to 155 liters (41 gallons). In the same period the obesity rate more than doubled, increasing by 112 percent.
- Between 1977 and 1994 the daily sugar intake per person rose from 235 calories to 318 calories, with soft drinks accounting for more of the increased sugar intake than fruit juices and desserts combined.
- Between 1970 and 1990, in conjunction with the rise in soft-drink intake, the amount of fructose (a very sweet sugar most often found in corn syrup) consumed per person increased 1,000 percent.
- Fructose stimulates the synthesis of fatty acids in a process called *lipogenesis*. It fails to trigger the production of insulin, thereby contributing to the onset of diabetes, as well as of the hormone leptin, which regulates appetite and satiety.
- Soft drinks increase appetite, inhibit satiety, and contribute to the consumer's desire for more sugar.
- Soft drinks have often overtaken and replaced milk and pure fruit juice as preferred beverages. This shift has resulted in a lower per capita intake of lactose, calcium, vitamins, and trace minerals, all of which the human body needs in order to grow and stay healthy.

In his investigations of the Coca-Cola phenomenon, journalist William Reymond presented similar conclusions in three separate books: *Coca-Cola, l'enquête interdite* (Coca-Cola, the Unauthorized Report), *Toxic*, and *Toxic Food*. He went further, however, and showed just how widespread the use of industrially produced glucose-fructose syrup is: up to 40 percent of all prepared foods, from soft drinks to fruit cocktails and nectars, including beverages and foods produced for babies, contain glucose-fructose syrup.

Consumption of this syrup is linked to the sharp rise in obesity and diabetes around the world, particularly among the disadvantaged, whose diets consist mainly of fast foods. In one impoverished community that William Reymond observed, Rio Grande City, a border city in southern Texas, 50 percent of ten-year-old children were obese and half the population suffered from diabetes.[6]

Iced Colas

A medical colleague recently traveled to Canada's Far North to visit aboriginal villages devastated by drug addiction and suicide. Upon his arrival in one community he noticed that the children had large quantities of soft drinks and hot dogs in their hands. Wondering if they were celebrating a birthday, he asked what the occasion was. To his amazement his host replied, "There's no special occasion. This is what they normally eat."

Does this refrain hold true for people with higher incomes and more education? Do they eat fast food every day, too? It seems likely that they do, according to information published in the 2008 statistical directory on food put out by Quebec's Ministry of Agriculture, Fisheries, and Food. It found that in North America almost half of all meals are eaten outside of the home. In restaurants at lunch hour, carbonated drinks are the second most frequently ordered item (22 percent) and, in the evening, the fourth most frequently ordered item (20 percent). What food was the first choice at both lunchtime and dinner? French fries!

There is no more frightening convergence of fattening foods than the infamous hamburger-fries-cola combo. Hamburgers, fries, and the condiments poured on them are overly salty. Eating heavily salted foods makes people thirsty, and so they reach for a carbonated drink. The glucose-fructose syrup in these drinks, as we shall see, inhibits the feeling of fullness (satiety), which in turn makes people eat more. In addition, drinking sodas, which release carbon dioxide, distends the stomach. Over time, those who chronically eat fast food must eat more in order to feel full.

More salt, more soft drinks. More physiological imbalance—and more obesity.

The news is especially bad for people who are obese. With diet alone, less than 5 percent will return to a normal weight. The only effective treatment for those suffering from severe obesity is bariatric surgery, which reduces the size of the stomach—a very drastic, even dangerous, means for them to achieve satiety.

The Link to Diabetes

The sudden and rapid growth of obesity in North America and around the world has led to a sharp increase in type 2 diabetes: between 1980 and 2005, the number of diabetics in the United States increased from 5.6 million to more than 15 million.[7] In Canada, the number of diabetics went from 700,000 in 1995 to more than 1.5 million in 2005.[8] Diabetes has become a global threat, with the potential to wipe out the last five decades of improvements in cardiovascular health that I described in an earlier book, *Prévenir l'infarctus ou y survivre* (Prevent a heart attack or survive it).[9]

Eighty percent of adult-onset diabetes cases are linked to excess weight. This simple fact should awaken public health authorities to the urgent need to fight the specific causes of obesity, especially soft drinks.

Another unpleasant surprise: certain foods and food additives help *induce* diabetes. There is a correlation between the increased intake of so-called refined sugars, mainly high-fructose corn syrup, and the alarming rise in the average body weight of North Americans. Today high-fructose corn syrup comprises 20 percent of all sugars consumed by North Americans. In other words, this one "refined" sugar contributes more to obesity and diabetes than any other carbohydrate in the diet.

The evidence is in. Soft drinks can directly bring about obesity,[10] diabetes,[11, 12] and high blood pressure,[13] and therefore heart disease.[14] To top it off, cola drinks also contain phosphoric acid, which contributes to osteoporosis and bone demineralization. A 2006 study at Tufts University in Boston showed a 3.7 percent reduction in bone mineral density at the femoral neck in women who drank one or more cola drinks per day.[15] The phosphoric acid in cola beverages disturbs the calcium-phosphorous equilibrium, the very basis of bone health.

The 2008 German study DONALD followed 228 children and concluded that cola consumption has a detrimental effect on bone mineralization in children. Colas contribute to decreases in bone mineral content, the surface area of cortical bone, and bone strength. These three indicators demonstrate an overall decline in the quality of bone modeling and remodeling in growing youths, which would explain the pronounced osteoporosis among cola drinkers.[16]

The "Purity" of Fruit Juice

Despite their level of natural sugar, pure fruit juices such as orange, grapefruit, and apple are not associated with the upsurge in the above conditions. In fact, they are strongly beneficial in keeping our bodies in good health.

It is important, however, to make a distinction in the type of fruit juice. There are pure juices—fresh, frozen, and reconstituted—and then there are "nectars," "cocktails," and "fruit drinks" in which the original juice is modified—in particular, by the addition of glucose-fructose syrup. Take, for example, cranberry juice. Praised in recent years for its antioxidant benefits, the cranberry has shot to stardom in the world of fruit. This tiny berry has the ability to prevent bacteria from growing on the wall of the bladder. Twenty years ago in Quebec, almost no one drank cranberry juice, except those suffering from chronic urinary tract infections, who drank it as a medication. But marketers soon recognized the berry's appeal among health-conscious consumers, and today cranberry juice and several other fruit and vegetable juices are top sellers. Although this trend is good overall, consumers must read juice labels carefully to know exactly what they are getting. Beverage companies have been known to tamper with the natural goodness of the cranberry, pumping the cranberry "cocktail" full of diabetes-inducing glucose-fructose.

Good Fat, Bad Fat. Good Sugar, Bad Sugar

In the simplest of terms, just as there are good fats (such as olive oil, a mainstay of the Mediterranean diet) and harmful fats (trans fats and an excess of saturated fats), there are good and bad sugars.

So how, physiologically, does industrially produced fructose promote obesity? A 2013 study at Yale University in New Haven, Connecticut, used the formidable power of magnetic resonance imaging (MRI) to examine whether the action of fructose in the brain was different from that of glucose.[17]

The researchers found that ingesting glucose reduced the flow of blood to the hypothalamus, the region of the brain responsible for the feelings of hunger and satiety. Glucose reduced the activation of the hypothalamus and stratium and thus increased feelings of satisfaction and fullness. In contrast, ingestion of fructose did not reduce the activation of the hypothalamus but rather mildly stimulated it, thereby reducing the feeling of fullness.

The study seemed to show that the key to losing weight is not simply to eat less food but especially to eat less of the foods that disrupt the satiety center—in particular, less of the fructose that is so prevalent in industrially produced foods.

Are "Diet" Soft Drinks a Healthier Choice?

The beverage industry developed diet drinks to appeal to consumers concerned about the amount of sugar they were taking in. In these drinks, artificial sweeteners such as aspartame replace the glucose-fructose syrup used in the original recipes. "Sugar-free" drinks, people believed, had none of the harmful properties of their sugary counterparts.

However, a French study undertaken by the National Institute of Health and Medical Research tracked 66,188 women for fourteen years and compared the health impacts of drinking diet soft drinks to those of drinking the original sugary drinks and drinking fruit juice. This massive study found that participants who drank the largest quantities of the "diet" or "light" drinks and those who drank the largest quantities of the "normal" soft drinks were, respectively, 2.2 and 1.34 times as likely to develop type 2 diabetes than the women who drank only natural fruit juice.[18] Once again we see that the road to hell is paved with good intentions!

Public programs aimed at controlling soft-drink consumption have had very little impact until recently. Almost everywhere you go, there are soft drinks for sale. These beverages are easy to produce and ship, have almost unlimited shelf life, and are very inexpensive to produce. Consequently, soft drinks are immensely profitable for the corporations that make them. So profitable, in fact, that the corporations can afford to pay image factories billions of dollars—for TV ads alone—to produce advertising campaigns aimed at convincing consumers to "open more happiness."

In 2010 nutritionists at the University of North Carolina published a study entitled CARDIA that had followed the dietary behavior of 5,115 participants for twenty years. It concluded that consumption of carbonated drinks would decline by 7 percent if the price of the drinks rose 10 percent. To the advocacy groups and others who had long called for higher taxes on soft drinks in order to discourage people from drinking them, here was much-needed ammunition.

Anticipating a backlash from all opposed to such a tax, especially the beverage industry, the editors of the *Archives of Internal Medicine* in which the CARDIA article appeared argued that the subsidies corn producers receive enable soft-drink manufacturers to set artificially low prices for beverages sweetened with glucose-fructose.[19] In other words, increased taxes on these drinks might level the playing field.

Whether or not higher taxes on soft drinks would discourage people from consuming them, as the tax advocates argue, citing the above study among others, what we really need to do is convince people to drink the many healthier and equally satisfying alternative beverages. Governments may be finally helping their young populations to make better choices. In 2005 California became the first American state to pass a law prohibiting the sale of soft drinks in schools. Philadelphia, Miami, Edmonton, and Vancouver are among the North American cities to ban the sale of soft drinks in school vending machines and food outlets. France and Great Britain have similar controls.

It takes 5 liters of water to produce 1 liter of soft drink—an ecological issue that we won't delve into here. Of greater concern is that, except for the water, all of the ingredients in a soft drink are harmful or of negligible nutritional value.

Which brings the discussion back to the Coca-Cola Company, this time to its campaign for the 2008 Olympic Games in Beijing. Wang Wei, former executive vice president of the Beijing Organizing Committee for the twenty-ninth Olympiad, stated that

> Coca-Cola was the most recognized and effective global sponsor of the Beijing Olympic Games. Through their comprehensive marketing programs allowing people from all over China to experience the Games, and by providing beverages to all the spectators and athletes, Coca-Cola's sponsorship made a significant contribution and helped bring the Olympic Movement to China.[20]

The company also stated, in all candor: "We also furnish the athletes with fruit juices and bottles of water."

How do cardiologists sum up the situation? We emphasize that the global explosion of obesity, diabetes, malnutrition, food deficiency, osteoporosis, and heart disease goes hand in hand with the sale of soft drinks and, in turn, with the advertising for them by fast-food corporations that are now targeting the huge markets of Russia, India, and China. "Open happiness" will prove to be a very bad beginning for them.

Obesity and Doctor Visits

At a national level, what impact does the obesity epidemic have on health services? In 2012 James McIntosh, a professor in the Department of Economics at Concordia University in Montreal, studied the effect of obesity on doctor visits in Canada. In his report for Health Canada, McIntosh found that without obesity, doctor visits would decrease by 10 percent nationwide. He further argued that the reduction in doctor visits would be even more pronounced if the survey had included the numerous consultations for problems related to type 2 diabetes, a disease that is 80 percent attributable to obesity.[21]

Happiness and optimism... Perhaps for the pharmaceutical industry?

six

SODIUM AND GOMORRAH

The Salt of Life

As indispensible and inescapable as they are in our lives and our kitchens, some ingredients are overly abundant in our food. Salt (sodium chloride or NaCl) is one such substance.

Salting is one of the simplest ways of preserving food and has been in use for thousands of years. Like the sugar in jam and the fat in pâté—indeed, like the cooking process itself—salt excels at inhibiting the growth of bacteria and mold.

There is growing international consensus, however, that the food industry has gone too far in its use of salt. Many of us now ingest two to three times as much sodium as we should. Although an adequate daily intake of salt is 1,500 milligrams, the average Canadian absorbs 3,400 milligrams, and a young adult male as much as 5,000 milligrams, of salt per day.[1]

The problem isn't table salt, which represents only 10 percent of all salt consumed, nor the salt that occurs naturally in foods, which accounts for another 15 percent of total salt consumption. The main problem is the generous amount of salt found in even sweet desserts, bread, breakfast cereals, so-called pure vegetable juices, and numerous other popular and easily accessible industrial foods. They are the source of 75 percent of the salt intake of Western populations.

The food industry adds salt to its "goods" to make them tastier and to extend their supermarket shelf life, thus reducing spoilage and boosting profit margins. The industry also creates eye-catching advertising campaigns to attract more buyers to the very products that aggravate consumers' overconsumption of salt.

Salt and Blood Pressure

Once again, our arteries are under attack. The Heart and Stroke Foundation of Canada reports that one in five Canadians suffers from hypertension and that three in ten develop the condition because they eat too much salt, often unknowingly and from foods in which they least expect to find it.

A recent study in the *New England Journal of Medicine* estimated that, in the United States, a reduction in the average daily intake of salt by 3 grams (or 1.2 grams of sodium) would reduce the annual number of heart attacks by 76,500, strokes by 49,000, and all deaths by 68,000. The annual savings in health-related costs from such reductions would be in the range of $10 billion to $24 billion. Furthermore, if the U.S. population could reduce its average salt intake by just 1 gram per day, the estimated benefit would be the same as treating every hypertensive American with medications.[2]

A cardiologist who just now learned that only 10 to 15 percent of dietary salt comes from the salt shaker and 75 percent comes from industrial foods may feel a twinge of regret for having poured guilt on patients with admonishments to "stay away from the salt shaker" and "eat a salt-free diet." The real problem all along was the groceries in patients' shopping carts and the restaurant meals they ate. Live and learn, especially in this field.

In the 1980s, medical schools taught aspiring doctors that high blood pressure was "essential" in up to 90 percent of cases; in other words, it was genetic. But later studies on the effects of salt on human health, on fine particulate matter from pollution (Chapter 9), on lead (Chapter 7), and on soft drinks (Chapter 5) emphasized the role of the environment among the causes of hypertension. As we recall from the Tsimane people (Chapter 2), hypertension is not inevitable.

Trans Fats. Them Again?

Other substances, like trans fats, simply have no good reason to exist. You might think that the message on the detrimental effects of trans fats in foods has been heard and that trans fats are no longer being used. That is only partially true, however, in developed countries and not even close to the actual situation in emerging countries. New York and a couple of other municipalities and countries have banned trans fats outright, and the debate on their safety is heating up internationally. Given the overwhelming data confirming the cardiovascular and metabolic hazards of trans fats, it is high time that trans fats be abolished.

The correlation between vascular disease and harmful cholesterol—low-density lipoprotein cholesterol (LDL-C)—has been well documented by the Framingham Heart Study, among others. Every 1 percent increase of LDL-C in the blood results in a 1 percent rise in cardiovascular disease. In Canada it is estimated that almost 40 percent of adults have high levels of "bad" cholesterol, as LDL-C is often called.

When this association became known, health authorities took the easy way out and blamed cholesterol in the diet. This assumption was misleading, however, because cholesterol is an essential component of our cells, particularly of neurons, or nerve cells.

Errors of this type abound. For a while people were told that "cholesterol plugs arteries." Even some doctors added to the confusion when, by trying to make the science easier to understand, they told their patients that heart attack and stroke are caused by "cholesterol plaques." This idea still circulates widely in the general population.

In fact it is the long, complex process of atherosclerosis, in which cholesterol plays only a supporting role, that narrows arteries—not cholesterol alone. And it is internal blood clots, not plaques, that suddenly block arteries completely, causing the disastrous tissue death that we call heart attack and stroke.

The cholesterol occurring naturally in our food rarely causes adverse health effects. This cholesterol represents 20 percent of the total cholesterol in the human body, while the liver synthesizes the remaining 80 percent.

More than any other factor, what stimulates our own production of cholesterol is dietary fat. In particular, *saturated* fats and *trans* fats raise our cholesterol levels, because these substances interfere with the liver's natural processes for regulating cholesterol.

It took some years for researchers to discover that one cause—and perhaps the *main* cause—of high blood cholesterol is the trans fats in industrially produced foods. These are much more harmful than the cholesterol that occurs naturally in food. For instance, consumption of trans fats increases LDL-C in healthy individuals just as much as it does in patients with the hereditary form of hypercholesterolemia (high cholesterol in the blood).

There are two types of trans fatty acids: natural (found in dairy products and red meat) and synthetic (found in hydrogenated vegetable oils and fats). The latter were developed in the early twentieth century and are among the some 120,000 synthetic substances that humans have introduced into the environment.

A chemical reaction known as partial catalytic hydrogenation can transform liquid oils into a solid or semi-solid state, thus creating margarine and shortening, for example. The texture of such products is often better suited to the needs of the food industry. Further, hydrogenated oils also help preserve any foods made with them because they oxidize and become rancid relatively slowly. Thus, restaurants and other food retailers find chemically transformed oils to be attractive and profitable.

In North America, synthetic trans fats are found in 40 percent of all industrial foods. They are used in French fries, microwave popcorn, croissants, pies, donuts, pastries, cookies, and all low-end chocolate products. They are even found in foods produced for babies and toddlers.

But—and this is a big but—trans fats are a leading cause of obesity, diabetes, hyperlipidemia, and cardiovascular disease. A person whose daily trans fat intake is 5 percent of his or her total fat intake has a 23 percent greater risk of cardiovascular death than someone who does not consume trans fat.[3] Canada had the highest levels of trans-fat consumption in the world in the 1950s and 1960s, years during which there was a corresponding peak in Canadian cardiovascular deaths.

Trans fats are associated with increased LDL-C, high triglycerides, and diabetes. They also reduce the body's ability to produce beneficial cholesterol, or HDL-C.[4] This discovery helps explain why, until very recently, significant discrepancies were found between the "normal" blood cholesterol levels of Americans aged fifty to seventy years and those of Asians of the same age. These so-called normal rates were much higher for North Americans because this group consumed more trans fats.

Unfortunately, the massive introduction of industrial foods into Asia in recent years has been accompanied by a sharp rise in the consumption of partially hydrogenated oil. As a direct consequence, the incidence of high cholesterol among Asians has skyrocketed. From New Delhi to Beijing, cardiovascular disease is now rampant.

Trans Fats and Public Health

Denmark has emerged as a world leader in the fight against synthetic trans fats. In 2003 it prohibited all oils and fats containing more than 2 percent industrially produced trans fatty acids. Even the largest multinational food corporations had to fall in line and reduce the proportion of trans fats in products sold in Denmark. Elsewhere, however, it has been business as usual.

Canada's reaction to the epidemic of trans fats in food has been tepid at best. In 2006, the Trans Fat Task Force co-chaired by Health Canada and the Heart and Stroke Foundation of Canada stated in its final report:

> The Danish approach—a 2 percent limit on industrially produced trans fat content of oils or fats used in foods—would not be the most appropriate course for Canada. A higher limit [4 to 5 percent], which included all sources of trans fat, would be more feasible to implement and could still yield a significant health benefit to the Canadian population.[5]

The task force's position that limiting the trans-fat content of foods in Canada to 5 percent would result in fewer cases of cardiovascular disease is nonetheless acknowledgement of how widespread and harmful trans fats are.

In June 2007 Canada's Minister of Health reacted to the recommendations of the final report of the task force:

> We are giving industry two years to reduce trans fats to the lowest levels possible, as recommended by the Trans Fat Task Force. If significant progress has not been made over the next two years, we will regulate to ensure the levels are met.[6]

Since 2006, when it became mandatory in Canada to indicate the presence of trans fats on the Nutrition Facts label, the consumption of these fats has declined among Canadians. Eighty percent of prepackaged foods respect the imposed limits. But foods sold in cafés and restaurants do not always comply: 75 percent of the croissants, donuts, and muffins sold in bakeries and restaurants exceed the limit of 5 percent trans fat. The Canadian Heart and Stroke Foundation reacted to this deplorable situation in January 2010:

> Some food companies have taken strides in eliminating artificial trans fats—while others have not. Unfortunately, there remain too many products that continue to contain disturbingly high amounts of these fats. This includes many foods often consumed by children such as cakes, donuts and brownies. The bakery sector in particular, including between 33 and 75 percent of some of these products, continues to be riddled with unnecessarily high levels of trans fats.
>
> The government's two-year voluntary period has come to an end. Trans fat levels are still too high, especially in baked goods. The verdict is in. Canada urgently needs trans fat regulations to protect our children and all Canadians. The time to move forward is now.[7]

McFrappé and McStatin

Such calls for reform of the regulatory environment have yet to be heeded by Health Canada. As of early November 2013 it continued to support a voluntary approach to eliminating synthetic trans fats from Canadians' diets. South of the border, meanwhile, change was brewing. On November 7 of that year the U.S. Food and Drug Administration announced a proposal to phase out trans fats in processed foods. Having concluded that synthetic trans fats are no longer "generally recognized as safe" for people to eat, the FDA estimated that its proposed ban "could prevent an additional 20,000 heart attacks and 7,000 deaths from heart disease each year" among Americans.[8]

Physicians have expanded the debate on synthetic trans fats and the health impacts of industrial foods in surprising ways.

In 2010, Melissa Walton-Shirley, cardiologist at T.J. Samson Community Hospital in Glasgow, Kentucky, posted a humorous but nonetheless serious health advisory on her blog: "The McDonald's Frappé: Warn your patients of this latest example of food-industry terrorism."[9]

When the McDonald's restaurant chain introduced its new iced coffee, the McCafé Frappé, Walton-Shirley was eager to try it. She, like so many others, enjoyed a good iced drink on a scorching hot day. But she was shocked by the ingredients in the small-size iced drink that McDonald's sold: in 354 milliliters (12 ounces) there were 450 calories, 20 grams of fat, 13 grams of saturated fat, 1 gram of trans fat, and 62 grams of carbohydrate.

In a small coffee.

She announced that she intended to warn her patients about "another food-industry gaffe. They [McDonald's] have missed a grand opportunity to provide a healthy dessert treat to our population and instead have planted yet another destructive 'bomb' on their daily menu."[10]

Walton-Shirley was upset that Jan Fields, president of McDonald's USA—at the time responsible for the nutrition of some 47 million customers worldwide each day (the figure is now 64 million)—had failed to ensure that her customers were eating healthy food. (Fields is still listed as one of *Forbes* magazine's 100 most powerful women.) Walton-Shirley was frustrated that a woman such as Fields, at the top of the corporate ladder and with the power to influence the global food supply, could be responsible for damage to the overall health of humankind.

Another novelty food item, dubbed "McStatins" by skeptics, was proposed by a group of British researchers in a study published in the *American Journal of Cardiology*.[11] Statins are a class of drugs that inhibit HMG-CoA reductase, a key enzyme in the synthesis of cholesterol. They are extremely effective in lowering blood LDL-C. Some common brand-name statins are Mevacor, Zocor, Lipitor, and Crestor.

The British study evaluated, in all seriousness, the benefits of adding a statin to condiments served in fast-food outlets in order to neutralize the metabolic damage of other menu items. The authors calculated that the consumption of a statin-containing condiment would compensate for the cardiovascular risk of eating a 200-gram (7-ounce) hamburger with cheese and a small milkshake containing 36 grams of fat, including 2.5 grams of trans fat.

The proposed condiments to which the statin would be added? Ketchup, mustard, salt, mayonnaise, and glucose-fructose syrup. Would you like a Crestor with that?

The authors went on to argue that harmful condiments are supplied free of charge in fast-food outlets, so the condiment containing the protective statin should also be free. As they wrote:

> It cannot therefore be reasonably argued on safety grounds that individuals should be free to choose to eat lipid-rich food but not be free to supplement it with a statin. It would be no more sensible than allowing individuals to drive without training or a license but at the same time restricting access to seatbelts and airbags.[12]

When it appeared, the study caused an uproar and much laconic comment, particularly in blogs on medical websites. It does, however, reflect a certain reality. Fast food and manufactured food are omnipresent in our society and are major sources of diabetes, hypertension, dyslipidemia, and cardiac disease.

Greenland and Nunavik

While completing a post-doctoral fellowship in epidemiology at Laval University in Quebec City, French researcher Émilie Counil compared the health of the Inuit populations of Nunavik and Greenland and made a compelling discovery. Speaking at the international conference Arctic Change 2008, she showed the extent to which the health status of these aboriginal peoples depends on decisions made by others in distant capital cities.[13, 14]

In Greenland—governed by Denmark—foods containing more than 2 percent trans fats have been banned since 2003. In Nunavik, a vast area in northern Quebec that stretches from the eastern shore of Hudson Bay to the Labrador border, the regulation and monitoring of the use of trans fats in foods are less strict.

Counil found that the blood levels of trans fats in the population of Nunavik are three times the levels in the population of Greenland.[15] Among the nine hundred adult males she surveyed from the Nunavik region, she discovered an increase in the obesity rate from 19 percent in 1992 to 28 percent in 2004, as well as elevated blood insulin levels, a sign of insulin resistance and a precursor of type 2 diabetes. Other metabolic syndromes were also present, all of which result in an increased risk of cardiovascular events.

Not content simply to record the health risks faced by the population of Nunavik, Counil decided to examine the foods available for sale in Inuit communities in the region. Among her observations:

- The shelves of the Northern Stores and Co-op Stores were filled with foods containing high levels of trans fats. One example, Pop-Secret microwave popcorn from Betty Crocker, contained 8 grams of trans fat per pouch.
- In the community of Akulivik, the Co-op Store sold a wide selection of industrially produced pastries. The children and others who delighted in eating them were unaware of the high trans-fat content of these foods.
- Young Inuit in the communities of northern Quebec drank up to 1.4 liters (49 ounces) of soft drinks and other sugar-sweetened beverages such as Gatorade every day, thereby amplifying the metabolic syndromes that affect them.[16]

Consider the parallels among the northern peoples of Nunavik and Greenland, the descendants of the Okinawans, and the Tsimane people. The two Inuit groups represented an opportunity to conduct an experiment in public health. They became laboratory subjects for decisions made in the offices of faraway health ministries. Convincingly, they show us how the health status of two populations of the same ethnic group evolved differently in different environments—that is, two separate regulatory regimes.

Happily, Counil did not stop there. On her recommendation, a team from the Faculty of Medicine at Laval University, in cooperation with the Makivik Corporation of northern Quebec, organized a feasibility study to determine the possibility of replacing foods high in trans fats with similar products containing fewer or no trans fats. Then, in collaboration with two food distributors that supply stores throughout Nunavik, a young student researcher in human nutrition, Valérie Blouin, drew up a list of food products sold in the region on the basis of their fat content. In addition to identifying products high in trans fats, the list highlighted similar products that contained better-quality fats.[17]

Nowadays, those of us working in environmental health must contend with not just new poisons but ancient ones as well.

seven

ON WINGS OF LEAD

LEAD HAS BEEN a great boon to cardiologists. Lead armor protects us daily, shielding us from X-rays and gamma rays. The modern-day knights of the fight against disease—radiation oncologists, radiologists, and interventional cardiologists—don aprons, eyeglasses, and collars impregnated with lead to prevent the leukemias, other cancers, and cataracts linked to radiation exposure. Even the walls and windows of radiology labs contain lead.

Like fire (see Chapter 8), lead has played a long and fascinating role in human history. It was the first "plastic," in the true meaning of that term, because it is pliable and easy to work. Lead is accessible, stable, and long lasting and has thousands of uses. It is an integral part of the history of humanity.

Forty thousand years ago, the first prehistoric tombs were decorated with paintings whose pigments contained lead. Six thousand years ago, people fashioned the first jewelry from lead, no doubt because lead can be worked into an infinity of shapes and is more durable than wood or bone.

Engineers of the Roman Empire designed the first system of aqueducts, which supplied constant fresh water to the capital and enabled the Roman public baths to be built. In fact, Rome sat atop a vast network of underground lead pipes. The water-carrying use of lead, or *plomb* in French, is recalled even today in the terms *plumbing* and *plumber.*

From the time of Julius Caesar through to the Middle Ages, aristocrats and high-ranking officials ate from dishes containing lead. They also wrote with it, using a lead cylinder that served as the precursor of the "lead" pencil.

The flamboyant windows of Romanesque and Gothic cathedrals are held together with thin strips of lead, which allow these stained-glass treasures to give forth their brilliant light, a victory for humanity and spirituality. Further, when lead is added to glass, the result is the finest crystal. Lead is an element of many paint pigments and is sometimes used in cosmetics—for example, in *khol*, the dark eye makeup worn by women (and some men) in the Middle East and India.

The invention of moveable lead type by Johannes Gutenberg set off a torrent of reading and writing that democratized communication around the world. Gun cartridges containing lead revolutionized warfare as well as hunting, for both food and sport. Even fishing was dependent on this soft metal, which was used to weigh down fishing lines.

In the industrial era, lead appeared everywhere: in oil paint and gasoline, batteries and dental fillings. With the discovery that the high atomic mass of lead allows it to block X-rays and gamma rays, lead protective gear was introduced into heart catheterization labs, radiology labs, radiotherapy labs, and dental clinics to protect patients and staff from harmful radiation.

But! Despite all the services that this metal has offered to humankind, lead has proved to be very dangerous to all who use it.

An Insidious Assassin

Several anthropologists attribute the fall of the Roman Empire, at least in part, to lead. Lead poisoning was called "saturnism," since the alchemist's sign for lead was the symbol for the planet Saturn. The Roman aristocracy lived immersed in lead: it was in their water, in their food, and even in their wine. Roman wine was preserved using lead acetate, which gave it a slightly sweet taste. The result was that the Roman aristocracy, amid their opulence and technological advances, little by little developed encephalopathy and dementia. (Even in Julius Caesar's time, it was very risky to put chemicals into food!)

Modern medicine describes the complications of lead poisoning as follows: neurological and cognitive deficits, hyperactivity with difficulties in concentration, memory loss, fatigue, confusion, erratic behavior, dementia, lethargy, coma, and death. Recently researchers have also established that lead also plays a role in the development of certain schizophrenias.

History repeats itself, and incidences of lead poisoning have occurred regularly over the centuries. We have seen that Roman nobility suffered from dementia related to lead poisoning. In the Middle Ages, German monks who drank wine sweetened with lead acetate or "sugar of lead" fell victim to "lead colic," the same illness that affected stained-glass workers and painters. During the two world wars, scores of workers in munitions factories suffered from exposure to lead.

Lead in munitions can still be a health threat, as people living near Lake Saint Pierre close to Trois-Rivières, Quebec, discovered. For almost half a century the lake served as a firing ground for Canadian army training exercises. Today, efforts are underway to de-mine the lake to remove the danger from unexploded shells and also to mitigate the alarmingly high level of lead in the water.

With regard to hunting, wild game killed with lead buckshot may be toxic, so people are told to remove the game meat surrounding the pellets. In urban centers, industrial paints and leaded gasoline have become major environmental toxins, compelling governments to try to eliminate them, though implementation rates vary greatly from one country to another.

Barbie in China

Despite lead's toxicity, which has been known since antiquity, and despite strict regulations on lead use today, the metal's versatility and affordability have led to many recurrences of lead poisoning. The American toy company Mattel was forced to implement a massive recall of some of its products for exactly this reason—an unanticipated side effect of globalization.

According to a CBC News report from September 5, 2007,

> The U.S. Consumer Product Safety Commission, in co-operation with Mattel Inc., announced late Tuesday that it is recalling about 675,000 Chinese-made toys that have excessive amounts of lead paint.
>
> The voluntary recall covers units of various Barbie accessory toys that were manufactured between Sept. 30, 2006, and Aug. 20, 2007.

"Consumers should stop using recalled products immediately unless otherwise instructed," said the agency's website.

As well, 8,900 different toys involving Big Big World 6-in-1 Bongo Band toys from the company's Fisher-Price brand were recalled. Those products were sold nationwide from July 2007 through August 2007.[1]

Controlling the Killer

Lead does not occur naturally in the human body, unlike a number of other metals, including iron, copper, and magnesium. Perhaps because of this fact, there is no universally accepted toxic threshold for lead. Standards have varied from country to country and from decade to decade. In France in 1976, the toxic threshold for lead in the blood was set at 400 micrograms per liter (µg/L); today, it is 50 µg/L. The World Health Organization and many countries set the acceptable threshold at 100 µg/L, but other experts claim that adverse effects on the brain appear below this level—indeed, at *any* level.

One interesting study showed that children born to mothers with high blood lead levels have a greater risk of mental illness.[2] In 2004, Ezra Susser of Columbia University in New York City followed 12,094 children born in Oakland, California, between 1959 and 1966. Blood samples collected from the mothers when they were pregnant were frozen and stored for future analysis. Susser subsequently found that, among children exposed to high lead levels *in utero*, incidents of schizophrenia and related disorders were twice as frequent.

Even more startling is the emerging evidence that criminal behavior is strongly correlated with the level of lead found in the environment.[3] In May 2000, Rick Nevin concluded that lead exposure may explain between 65 and 90 percent of the variation in the rates of violent crime in the United States, with rises in crime systematically linked to rises in the average concentration of lead in the blood. According to Nevin, the gradual removal of lead from the environment has been accompanied by a decline in crime, mainly owing to the lower incidence of cerebral lead poisoning and its associated violent behavior.

Although humans have known about lead poisoning since antiquity, we are only now beginning to understand the relationship between lead exposure and cardiovascular health.

Ironically, lead remains beneficial and protective for interventional cardiologists when it is incorporated into the kind of protective gear that prevents the leukemia that killed Marie Curie, who handled radioactive radium without any protection. However, there is now ample proof that inhaling and ingesting the lead in our environment has resulted in serious public health problems. Although public authorities have been taking action to remove lead from gasoline and paint—and strict regulatory measures are essential—we must still face the technological challenge of finding safe replacements for lead.

A Chilling Wake-up Call

The stakes are high indeed. Leaded gasoline has poisoned the atmosphere. Glacial ice-core samples from Greenland reveal that lead levels in the ice have increased by some two-hundred times since the beginning of the industrial era. A Chinese study that examined ice cores from the Dasuopu glacier in the Qinghai-Tibetan Plateau showed that between 1953 and 1996 the concentration of lead in the ice rose steadily to a level twenty times its initial level.[4]

We now know beyond a doubt that the glaciers are melting. But we don't know what will happen to the meltwater in its downstream basins, some of which carry great rivers through densely populated areas. Will a century's worth of lead accumulation in glacial ice be transported away by meltwaters in just a few short years—and contaminate the water supply of a vast number of people?

This brief side tour into the role of lead in human history is but an introduction to a serious environmental issue related to heart disease. Humans have made great advances in science and culture, but some discoveries, including that of lead, have had harmful effects that may be hard to reduce and even harder to eradicate.

Another case in point is fire, perhaps our most important discovery of all. The domestication of fire has contributed to the advancement of humanity in innumerable ways, but today fire is also a major cause of climate change and the dramatic rise in heart disease globally. To find the culprit we need look no further than the ubiquitous burning of fossil fuels.

eight

PLAYING WITH FIRE

APPROXIMATELY 2,500 YEARS AGO, classical Greek thinkers laid the foundations of modern Western science and philosophy. One of the builders of this scientific legacy, in which trigonometry intermingles with mythology, was Empedocles (490–430 BC), who lived in Agrigentum, Sicily. It was Empedocles who declared fire, water, earth, and air to be the four essential components of physical and spiritual life. More recently this ancient and seductive paradigm has inspired everyone from New Agers and rock musicians to novelists and fantasy-film directors.

Scientists, however, have searched beyond the symbolic quartet of fire, water, earth, and air to discover the actual four elements that support life: carbon, hydrogen, oxygen, and nitrogen, or CHON. These elements are fundamental to all living things.

In recent years former U.S. Vice President Al Gore has become something of a modern-day Empedocles. Through his books and films and his vision for the environment, he revisits a worldview based on the elements of earth, air, fire, and water and shows how important it has been to our lives by drawing from a variety of expressions from different cultures, including a moving and poetic Sikh proverb: "Earth teaches us patience and love; Air teaches us movement and liberty; Fire teaches us warmth and courage; Water teaches us purity and cleanliness; and Sky teaches us equality and tolerance."[1]

From this idealistic view, Gore moves on to deliver the bad news—the "inconvenient truth." He shows how Earth's vital elements are under attack and rapidly being degraded, and he explains how the damage to earth, air, and water is causing devastating climate change. One of the most revealing images of this damage is reproduced in Figure 1. The carbon dioxide curve depicts the alarming rise in the concentration of CO_2 in the Earth's atmosphere as measured at Mauna Loa Observatory in Hawaii. Clearly, something is off kilter in this image. The failure to maintain the millennia-old equilibrium of nature, and the need for a return to it, is also the subject of scientist David Suzuki's *The Sacred Balance.*

Not only is this image central to the thinking of Al Gore and other respected environmentalists, including the scientists of

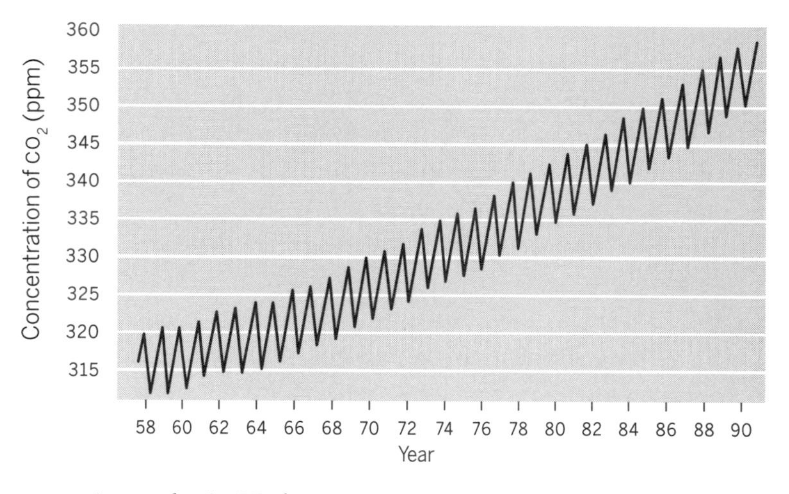

FIG 1 *Atmospheric CO_2 from 1968 to 1991 at Mauna Loa Observatory.*

the Intergovernmental Panel on Climate Change (IPCC); it is also crucial to our understanding of the planetary debacle heading our way. But the scientists have more bad news: atmospheric CO_2 is increasing and is doing so at a faster rate than at any time in Earth's history. At the same time, the average global temperature is also rising, increasing roughly 0.85 degrees Celcius (1.53 degrees Fahrenheit) between 1880 and 2012, according to the Intergovernmental Panel on Climate Change.[2] In May 2012 the National Oceanic and Atmospheric Administration (NOAA) reported that the concentration of atmospheric CO_2 at Barrow, Alaska, reached 400 parts per million (ppm). In 1850, when the industrial revolution began gaining momentum in North America, the average global level was 280 ppm. Indeed, atmospheric CO_2 is now at its highest level in 800,000 years.[3]

In 2009 researchers from thirteen governmental scientific agencies, major universities, and leading research institutes, including the Lawrence Livermore National Laboratory, published *Global Climate Change Impacts in the United States*, a study that corroborates the meteorological models of the IPCC on the rise in the Earth's average temperature. According to the authors, the primary cause of global warming is human activity, in particular the activities responsible for the massive—and growing—production of greenhouse gases.[4]

Ashes in the Human Footprint

And what is at the heart of this global crisis? The primary historical cause of the increase in atmospheric CO_2, and thus global warming, is *the human discovery of fire*.

Before hominids conquered fire around 400,000 years ago, accidental fires in nature occurred infrequently. They flared up in twigs or leaves during dry spells or unusually hot weather, when lightning struck dry timber, or when sparks flew out of a volcano and landed on nearby vegetation. Such conflagrations laid waste to every living thing in their paths. Although in the short term each one was catastrophic, forest fires and volcanic eruptions proved to be sources of renewal and new life over the long term. You need only to dive among the magnificent coral reefs of the West Indies or walk through the rich variety of plants on La Grande Soufrière volcano in Guadeloupe to appreciate the great beauty and diversity of life generated by volcanoes. During the millions of years before the industrial era, however, fires in nature were rare and affected only limited areas. The cycle of life on Earth kept oxygen and carbon dioxide in a natural and effortless equilibrium.

It wasn't that way in the beginning. The very high level of atmospheric carbon dioxide that existed when Earth was formed gradually diminished when plant life appeared. The process began with cyanobacteria, or blue-green bacteria, which used carbon dioxide to grow and then released oxygen into the atmosphere. Over hundreds of millions of years, these bacteria, together with photosynthesizing algae and plants, managed to transform much of the world's carbon dioxide into organic deposits such as coral, peat, coal, petroleum, and natural gas. The last four of these are called *fossil fuels*, because their carbon was laid down long ago and cannot be replaced during human history.

Humans appeared and began making fires during the last few moments of geologic time. With the coming of the industrial revolution—only two hundred years ago—they moved away from wood and began burning fossil fuels, releasing huge amounts of trapped ancient carbon back into the atmosphere. The more we humans burned, the greater the increase in CO_2, until CO_2 once again started to accumulate in the atmosphere.

More atmospheric CO_2 is the common denominator of both climate change and recent disruptions in the world's oceans. In the latter case, the oceans are becoming more acidic as they absorb more CO_2. This acid imbalance has already done great damage to coral reefs and many other forms of marine life.

If we think about it, of Empedocles's four elements, fire is the outsider. Fire is the source of disequilibrium in the three other elements. Air (N_2, O_2, and CO_2), water (H_2O), and earth (C, H, O, N, and trace elements) are key components of the mixture that creates life. Fire is an external natural force, like gravity or lightning.

The Quest for Fire

In the beginning, by gathering coals from spontaneous fires, then through trial and error, and finally as a result of experimentation and thermodynamics, the human species conquered fire for light, heat, mechanical power, and chemical transformation. Fire became humankind's foremost ally against cold, darkness, hunger, predators, and enemies. It represented a wondrous new source of energy.

Fire is eternally fascinating: the candle-lit dinner, campfire, woodstove, holiday bonfire, fireworks, artillery fire, missile fire, volcanic eruption, nuclear explosion.

It is intriguing to contemplate the exact moment when humans first tamed fire. In 1981, French director Jean-Jacques Annaud absorbingly depicted this moment in the film *Quest for Fire*. In 2007, the Musée de paléontologie humaine de Terra Amata (Terra Amata museum of human paleontology) in Nice, in southern France, presented a show on the same subject. According to the museum's website:

> The mastery of fire represents the most important milestone in the very long evolution of humanity. Long after having carved the first tools out of stone, humans took a definitive step beyond the animal stage by domesticating a natural and reliable energy source—fire. In addition to its implications for the techniques and styles of food preparation, the conquest of fire had radical consequences for the evolution of the spirit of prehistoric man.[5]

Fire immeasurably improved the daily lives and survival rate of cave dwellers and led to a deep transformation of the human species. The descendants of those cave dwellers learned to harness the power of fire and use this vast energy source to fulfill their needs and desires. And then, many thousands of years later, in July 1969, after a three-day journey in a rocket propelled by a fire of oxygen and hydrogen, the *Apollo 11* astronaut Neil Armstrong stepped onto the moon and spoke his famous words, "That's one small step for man, one giant leap for mankind."

For thousands of years fire was believed to be divine; it played an important role in all religions and in countless sacred rituals. A flame burned at all times in Greek temples. Ancient Romans practiced fire worship, and they charged young girls known as Vestal Virgins with keeping temple fires lit in the temple of Vesta, goddess of the hearth. Even today the lighting of a flame often signals a sacred ritual in religious ceremonies. And, of course, people the world over recognize the Olympic flame as a two-thousand-year-old link between antiquity and the era of the modern Olympic Games.

The Greeks tell the story of fire in the myth of Prometheus. The Titan Prometheus stole fire from Mount Olympus in order to give it to humans, who were eager to acquire this new power. Zeus was furious about the theft and sentenced Prometheus to a terrible ordeal: to be chained atop Mount Caucasus, where an eagle would come each day to peck at his liver, which would then regenerate so that Prometheus would have to endure his fate for eternity.[6]

In the face of today's ecological threats, the myth of Prometheus has been revived. It is now coupled with the story of Pandora's box, which, when opened, released all the ills of humankind.

Fighting Fire

Combustion is the root cause of the increase in greenhouse gases and of global warming. It is time for a simple order to be spoken: stop the burning. Find alternatives to combustion. Remove carbon from the energy equation.

If the conquest of fire occurred today, the feat would warrant Nobel Prizes in physics, chemistry, medicine, and peace, so crucial is it to the survival of humanity. But the first human beings to seek warmth and other comforts beside a bright flame could hardly have imagined that their numbers would eventually grow to seven billion people. Further, almost all these people burn gas, kerosene, oil, heating oil, wood, and coal for their heat, light, and energy to live and work, to think and play.

All of this individual burning has a collective impact greater than that of the largest volcanic eruption or the most widespread forest fire in the history of the planet. It is equivalent to the force of an asteroid smashing into Earth. Although the full consequences of such an event are hard to foresee, they would certainly be devastating and potentially lethal for this planet and the future of life on it.

So great has been the increase in combustion by humans that the smoke from it now threatens habitats the world over. Since life emerged on the planet and the atmosphere stabilized, the only other causes of very rapid rises in carbon dioxide besides humans have been natural disasters like massive volcanic eruptions and meteorite impacts.

According to paleontologists, Earth has already experienced five periods of mass extinction resulting from such natural disasters. The most recent took place 65 million years ago and led to the disappearance of the dinosaurs, a loss, which though regrettable, had a positive outcome for the human species. The extinction of these reptiles meant that humans faced no competitors in their plundering of Earth's biodiversity.

The dinosaurs and numerous other life-forms died off after an enormous asteroid struck the Yucatan peninsula, in what is now Mexico. This asteroid, estimated at 10 kilometers (6 miles) in diameter, hit with a force the equivalent of 100 trillion tons of dynamite, or a force two million times as powerful as the most powerful nuclear bomb. The impact carved a crater in the Yucatan that measures, inside the rim, 160 kilometers (100 miles) across and, on the outer edge of the rim, 300 kilometers (186 miles) across, with half of the crater on land and the other half underwater in the Gulf of Mexico.

The fires from the impact spread across the planet and released devastating amounts of carbon dioxide, which, in turn, caused a mass extinction of life-forms. An estimated 45 percent of marine species living near the surface of the ocean were wiped out, along with 20 percent of deepwater species, 15 percent of freshwater species, and 20 percent of terrestrial species.

The Human Volcano

Atmospheric carbon dioxide is rising precipitously, at a faster rate than at any other time in human history, such that the planet is at risk of a sixth extinction.[7] Why? Because of Prometheus's gift of the fourth element: fire.

Despite what we know about the environmental risks of burning fuels, humans have not found efficient alternatives to it. In the nineteenth century, when people burned whale oil for light and heat, whales were hunted almost to extinction. Happily for the whales, we no longer use their oil. Instead we use petroleum, which is also being overexploited: almost half of the world's oil reserves have already gone up in smoke, and we are projected to deplete the resource in another fifty years.

One secondary consequence of burning fossil fuels is extreme weather; each new year seems to bring with it weather events even more destructive than those of the 1990s. Yet meeting the challenges of climate change is likely to require an even higher energy consumption. The more CO_2 is released, the greater the environmental stresses from climate change; the more we heat and cool our homes, the greater our need for coal-fired and gas-turbine electrical power plants, which release yet more CO_2. In his book *La légende de la vie* (The story of life), Albert Jacquard reminds us just how vicious the cycle is:

> Human beings discovered the existence of petroleum reserves; they drilled, cranked open the valves, the oil flowed, and they burned it. Sometimes to warm themselves or to deliver their goods, but often to enjoy stupid circus games during which, like sheep suffering from "whirling disease," they turn in circles for hours, as fast as possible... In less than a century all of the oil will be gone. It is high time we realized that it doesn't belong to us. It belongs to the men and women who will come after us.[8]

Humans have changed the environment so profoundly that the biodiversity of living species, both terrestrial and marine, and ultimately the survival of humanity itself, is at risk. One mammal species in four, one bird species in eight, one-third of all amphibian species, and 70 percent of all plants may perish in the sixth extinction, according to the International Union for the Conservation of Nature.[9]

Is it still possible to halt the disappearance of various species, given the threatened increase in the world's population to 9.5 billion by 2050? The American biologists Paul Ehrlich and Robert Pringle of Stanford University think it is, but only if several radical steps are taken at the global level. They outlined these steps in the August 2008 issue of the journal *Proceedings of the National Academy of Sciences*, an issue dedicated to the subject of the sixth extinction.[10]

Ehrlich and Pringle unhesitatingly declared that "the fate of biological diversity for the next ten million years will almost certainly be determined during the next 50 or 100 years by the activities of a single species. That species, *Homo sapiens*, is only 200,000 years old."

Given that mammalian species—and we are mammals—typically exist for about one million years, *Homo sapiens* is in mid-adolescence, according to the researchers. "This is a fitting coincidence, because *Homo sapiens* is now behaving in ways reminiscent of a spoiled teenager. Narcissistic and presupposing our own immortality, we mistreat the ecosystems that produced us and support us, mindless of the consequences."[11]

What do paleontologists say about the relationship between *Homo sapiens* of the twenty-first century and fire, the gift stolen from the gods?

> During the Upper Paleolithic Age, if not before, human beings developed strategies to build fires. In recent times the factory production of lighters and matches has made those strategies anachronistic. The techniques and objects we now use to build fires are the product of a long, mostly forgotten history that involved trial and error, experimentation, observation, and discovery.[12]

The Sacred Fire of Medicine

Fire is central to the history of medicine. We boil water to purify it. We cook meat and other foods to help preserve them and to kill parasites and prevent food poisoning. At one time leprosariums were cleansed with flaming torches. Villages ravaged by epidemics were burned to the ground. The bodies of plague victims were incinerated to prevent transmission of the infectious disease to others. Wounds were cauterized with red-hot iron rods pulled from a fire. People inhaled healing vapors warmed by a flame, and medical and scientific instruments were sterilized in a flame.

In modern medicine, fire has largely been replaced by technologies that are more efficient at disinfecting, sterilizing, cleaning, and healing. The last vestige of the era of fire is the Bunsen burner, whose flame is still used to sterilize curettes in microbiology laboratories. This practice will soon be archaic, however; electric sterilizers are replacing the gas burner in most hospitals. The only remaining medical use of fire will then be incineration, which hospitals are still obliged to perform for sanitation, as no efficient alternative to the incinerator has yet been found.

Across Canada, various public programs have been introduced to urge people to reduce their use of fossil fuels, and thus the production of greenhouse gases, by changing lifestyles and modes of transportation. Hospitals have scrutinized common practices and protocols for years with the purpose of eliminating the direct burning of fossil fuels, and fire is no longer part of our medical arsenal. New and better technologies have taken its place. Surely other sectors of society are also able to find more efficient and lower-cost alternatives to burning fossil fuels?

Putting Out Fires

Does it make sense to eliminate the combustion that goes on everywhere on this planet? If we accept that Prometheus's flame is the source of our problems, it seems all we have to do to prevent continued environmental degradation is to eliminate combustion and restore the balance between oxygen and carbon dioxide. We must literally and figuratively put out fires and find a solution to the crisis if we are to reduce the production of greenhouse gases and slow climate change. This imperative, of course, applies worldwide.

Where to Beat Back the Fires?

The spark of gasoline engines currently sets in motion approximately half a billion vehicles around the world, and at the rate China and India are buying them up, this figure will soon reach one billion. Oil-burning furnaces heat our homes. Coal-burning power plants are the source of half of the electricity produced in North America and the main source of power in Russia. Symbolically speaking, lighting up a cigarette—or a hit of crack, marijuana, or cocaine—is not essential to life but is rather a cultural adaptation. We know how dangerous and toxic these practices are. But what we have to remember is that inhaling "exhaust" from all combustion is toxic, be it smoke from cigarettes, crack, coal-fired power plants, or gasoline engines.

When discussing our environmental problems, we are quick to blame the automobile. But cars are not the main source of the megatons of CO_2 and pollution that escape into the atmosphere; it is, rather, the internal combustion engine. If people drove nothing but hybrid or electric vehicles, the emission standards set by the Kyoto Protocol would be achievable. The average hybrid electric vehicle gets 30 percent of its horsepower from rechargeable batteries and emits 3.5 tons of CO_2 annually, compared with 5 tons for the standard gasoline-powered vehicle. Choosing to drive a hybrid electric car is a simple way for individuals to significantly reduce their contribution to greenhouse gases and airborne pollutants. These vehicles also produce less noise, a significant irritant and stressor in the urban environment.

The higher cost of hybrid vehicles has deterred many potential buyers, but the cost difference will likely disappear as production volumes rise. As of October 2012, Toyota had sold 4.6 million hybrid electric vehicles worldwide. Although financial and other incentives to buy hybrid vehicles exist in some countries, they have not been implemented globally. In many parts of Europe, for example, purchase subsidies and discounts in parking lots for downtown street parking encourage the use of low-emission vehicles. North America, meanwhile, drags its feet on the issue.

It is not far-fetched to predict that hybrid electric vehicles will soon achieve an electricity-gasoline ratio of 95 percent to 5 percent, with the gas engine acting as a simple generator that consumes 1 liter of fuel for every 1,000 kilometers (2,350 miles per gallon). This back-up engine assures the vehicle greater autonomy than is now possible with a fully electric-powered vehicle. In terms of power, electric motors have always offered greater torque than internal combustion engines.

Even Formula One racing is committed to reducing its carbon footprint. With a regulatory overhaul planned to take effect in

2014, racing fans may one day see hybrids or even all-electric race cars in the Grand Prix. This move would certainly boost the development and promotion of renewable-energy road cars and be a twenty-first-century milestone.

A "cardio-active" redesign of cities (see Chapter 13) would also radically change the relationship between humans and their vehicles. It is difficult, and indeed a hassle, to drive a car in Geneva, Stockholm, and Paris—all cities with comfortable and energy-efficient transportation alternatives.

If we wish to re-establish Earth's natural balance and the conditions that allow living organisms to flourish, the simplest way is to progressively (but urgently) douse the flames that humans have fanned ever since the discovery of fire. Turning back the clock on the industrial advances of the last century, we must slowly limit the burning of fossil fuels planet-wide by vigorously promoting the use of combustion-free energy. Flameless technologies must become the order of the day.

Compared with the billions of years remaining in the lifespan of the sun—the supreme keeper of the O_2-CO_2 cycle of life—Earth's remaining oil reserves will last for about fifty more years and the coal seams for 200 to 300 years.[13] Oil and other fossil fuels are derived from plant matter that required millions of years to be transformed. The process of generating new fuels will simply take too long; it cannot be counted on to feed our machines.

However, compared with the fossilized remains of plants, living plant matter represents an inexhaustible energy source. Plants capture CO_2 and release O_2, the opposite of what happens when fossil fuels are burned.

Energy and Biodiversity

A modern Empedocles might define energy as follows:

earth = geothermal energy
water = hydro energy

air = wind energy
fire = solar energy

These flameless energy sources, beyond helping to restore the O_2-CO_2 balance, have another distinct advantage that I, as an interventional cardiologist, am particularly aware of: the use of energy sources other than fossil fuels will, even in the short term, reduce pollution and smog and alleviate the harm done to the cardiovascular system.

Pollution in city centers is already killing people. In Montreal, pollution kills two thousand people every year; across Canada, twenty thousand; and in China, two million. These figures are conservative, as they take into account only the acute toxicity of periods of elevated smog and not the chronic effects of air pollution on the cardiovascular system. In 2003 the record heat wave in Europe and accompanying deadly smog caused seventy thousand fatalities. This tragedy was a warning shot for people everywhere. As well, the anticipated increase in extreme heat and smog events will have ever-more-serious consequences on health care systems.

According to the Intergovernmental Panel on Climate Change (IPCC), the United Nations agency dedicated to "the preparation of comprehensive assessment reports about the state of scientific, technical and socioeconomic knowledge on climate change, its causes, potential impacts and response strategies," the steadily increasing levels of air pollution and smog will lead to a corresponding rise in cardiovascular deaths. Air pollution must be added to the list of other uncontrolled risk factors for heart disease, such as a sedentary lifestyle and obesity. We must therefore take action within, at most, a few years or decades.

In the medium term, burning less will slow down or even reverse the decline in biodiversity, allowing us to protect the countless proteins of medical interest found in nature. Some researchers believe that the majority of naturally occurring medicinal molecules are yet to be identified. Don't kill the goose that lays the golden eggs, they say, since undiscovered therapies could offer great promise to both physicians and patients.

Finally, and taking the long view, we must avoid at all cost the Armageddon of a sixth extinction that would lead to the disappearance of 90 percent of Earth's plant and animal species by around 2100: in less than one hundred years! According to Edward O. Wilson, biologist and author of *The Diversity of Life* and *On Human Nature*, for which he was awarded the 1979 Pulitzer Prize for non-fiction, the emergence of human beings led to the disappearance of 10 to 20 percent of all other species. An estimated 30,000 to 50,000 species continue to be lost each year. We may not have crossed paths with the dinosaurs, victims of the fifth extinction, but we know the potential victims of the sixth extinction—they are our children and grandchildren.

Protection for the Atmosphere

Here's an interesting parallel: Because of the success of public programs to promote smoke-free environments and smoking cessation, human health has, in general, improved measurably. Hospital admissions for heart attacks have fallen. But smoke is not just contained within walls. At one time open fires burned inside igloos, teepees, yurts, and longhouses. Choking on the smoke, humans learned to direct it outside via chimneys, thus displacing the problem. In London, the Great Smog of 1952 killed 12,000 people in the city center (see Chapter 9). In 1956 the United Kingdom passed the Clean Air Act, which reduced air pollution through restrictions on burning coal in cities and towns.

Now we discover that pollutants have saturated the atmosphere, the thin layer of air covering our planet. Drawing on the words of the late Carl Sagan, David Suzuki describes how thin the atmosphere is: "If the Earth were shrunk to the size of a basketball, the atmosphere that we all depend on for our very survival would be thinner than a layer of varnish. That's it—and everything that our tailpipes, chimneys, and engines vent goes into that thin layer."[14]

From her viewpoint in the International Space Station (ISS), astronaut Julie Payette marveled at just how thin and vulnerable the layer of air was that surrounds the Earth. During her 1999 voyage, according to Edward O'Connor, author of *Julie Payette: Canadian Astronaut*, Payette and the other crew members experienced itchy eyes, headaches, and nausea—symptoms often associated with sick building syndrome—because of the high levels of CO_2 in the air of the ISS. From time to time the crew had to return to the space shuttle to recuperate. At least they had a life raft to go to. Unfortunately, if Earth's atmosphere ever experiences similarly high CO_2 concentrations, there will be no life rafts for us, no place to hide while the carbonous gases dissipate on the mother ship.

Experts in the American Academy of Sciences predict that humans will disappear from the surface of the Earth around 2200. If they are right, let us hope that nature will choose a dominant species that is more integrated with the natural world than we are: a species that will satisfy its aspirations and attain a quality of life in symbiosis with the environment that allows it to flourish.

The dinosaurs were the largest living beings to disappear in the fifth extinction. They were neither the cause of the cataclysm, nor were they aware of the forces that killed them.

The discovery of fire was a milestone in human evolution, and its energy became indispensable to the survival of the species. Paradoxically, however, fire may also be paving the road to the sixth extinction and to the horrors once visited on Prometheus.

nine

BURNING HEARTS

WHAT DO THE ENGINE of a car and a heart have in common? Four valves and an exhaust. The valves in each are similar, but the exhaust is entirely different: burned gasoline for the engine and ventricular blood for the heart. And the exhaust from one is killing the other.

Heart disease is a fascinating subject. We now know that cardiovascular disease (CVD) is rare in animals; CVD was not common in humans before the advent of the industrial revolution; and CVD is rarely found in humans who live outside of the industrialized world.

To briefly recap its evolution, in the first half of the twentieth century in North America, the incidence of CVD skyrocketed, tripling and even quadrupling. So alarming were the numbers that health authorities began investing millions of dollars in massive studies such as the Framingham Heart Study, which originated in

Massachusetts, and the InterHeart Study, which surveyed fifty-two countries.[1] Since then, cardiovascular deaths have gradually decreased in many parts of the world. After peaking across North America in the 1950s, for example, the death rate from cardiovascular disease in Canada fell from 370 per 100,000 inhabitants in 1981 to 138.6 per 100,000 in 2007.[2] According to medical journal editors, half of the decline was attributable to improvements in medical treatment. (This news was music to the ears of cardiologists.) The other half of the decline was due to the improved management of risk factors through preventive medicine. Salim Yusuf, of McMaster University in Hamilton, Ontario, and principal researcher on the InterHeart Study, summarized these achievements in a 2002 editorial, signed by the world's leading experts on cardiac risk factors, in *The Lancet*:

> In the mid-1950s, myocardial infarction and strokes (vascular events) were not considered to be preventable. This view persisted until the early 1980s. Over the past two decades, reliable data have emerged indicating that smoking cessation, ß-blockers, antiplatelet agents, inhibitors of angiotensin-converting enzyme (ACE), and lipid-lowering agents, each reduce the risk of vascular events to a moderate but important degree.[3]

Additional evidence of the solid clinical and epidemiological gains in the fight against heart disease was to be found in the data on life expectancy. As reported in 2002 in the United States, life expectancy had increased by five-and-a-half years over thirty years, with the decline in the cardiovascular death rate accounting for two-thirds of this improvement. (Cardiologists were overjoyed.) The theories postulated in the Framingham Heart Study had been validated, and the massive financial investment had been justified. Physicians and politicians alike celebrated the good news on June 6, 2002, when Claude Lenfant, director of the National Heart, Lung and Blood Institute, delivered a speech to the U.S. House Committee on Energy and Commerce Subcommittee on Health:

> Stroke death rates have plummeted, due in great measure to improvements in detection and treatment of high blood pressure. The average American can expect to live 5½ years longer today than was the case 30 years ago, and nearly 4 years of that gain in life expectancy can be attributed to our progress against cardiovascular diseases.[4]

Although it would be interesting to know why the House Committee on Energy and Commerce had a subcommittee on health, the most startling aspect of the 2002 session at which Lenfant spoke was that not one word was said about air pollution. Another two years would pass before the American Heart Association issued its first official warning about the impact of smog on heart health.

Moreover, the news is not all good. The nations of the former Soviet Union have cardiovascular death rates up to *ten times* those in Western Europe, according to a 2008 WHO report: 762.8 deaths per 100,000 inhabitants in the Russian Federation versus 68 per 100,000 in France among men 25 to 64 years old.[5]

In China, the cardiovascular disease rate has risen fourfold in the past fifty years. The phenomenon, the Asian version of the American CVD epidemic of 1950, is being fueled by China's industrial revolution.[6] The rate of cardiovascular deaths in China, which is still rising, now exceeds that in the United States, where it is falling; a 2005 study found that stroke accounted for 171 deaths per 100,000 people in China compared with 23 per 100,000 in the U.S. and that heart disease caused 159 deaths per 100,000 inhabitants in China compared with 150 deaths per 100,000 inhabitants in the U.S.[7]

Called upon to explain the pandemic of CVD in the developing countries, medical researchers dredge up the classic risk factors—hypertension, diabetes, obesity, and smoking—to the riveted attention of pharmaceutical companies and catheter manufacturers, ever conscious that the size of this new market is in the billions.

But the classic risk factors do not explain all of the fluctuations in the incidence of CVD worldwide since the end of the nineteenth century.

What Framingham Did Not Say

Despite the wealth of information that it amassed, the Framingham Heart Study failed to explain the devastating rise in CVD deaths in the United States in the first half of the twentieth century. Genetically, Americans of 1880 and those of 1950 were not all that different. So what had changed?

Let us, without too great a leap, hypothesize that the buildup of grime in people's arteries—and the associated heart disease, heart attacks, strokes, and deaths—were caused, to a significant degree, by air pollution. The cardio-environmental model would perhaps explain several observations—for example, the parallel between the rise (followed by a gradual decline after 1970) in airborne pollutants such as carbon monoxide and nitrogen dioxide[8] and that of cardiovascular disease.[9]

Looking back at the sharp rise in cardiovascular mortality in the first half of the twentieth century, we should perhaps consider the contribution of the two world wars—the 1939–45 conflict in particular. Hitherto unknown quantities of pollutants were generated during the Second World War because every ship, plane, and tank; every iron and steel mill; and every munitions factory was powered by a fossil fuel—primarily coal. Air quality was definitely not a priority during wartime. In addition, tons of pollutants rose into the atmosphere over Europe and Asia from the bombed- and burned-out cities and industrial and petroleum installations.

In the United States and in Canada, heavy industry operated at full bore before and after the war. When it was over, the United

States accounted for half of the world's GDP while Europe, Asia, and Africa were left to rise from ruins that today we might be able to imagine by looking at images of post-Katrina destruction and then multiplying that by several Exxon Valdezes and Fukushimas. Then, a few years after the end of World War II, American public health authorities sounded the alarm on heart disease, setting in motion the Framingham Heart Study.

The cardio-environmental model would also explain the inverse corollary of a paradise lost: the virtual absence of arterial disease and hypertension among the Tsimane people and others living beyond the reach of industrialization, its airborne pollutants, and its industrial food additives.

More evidence of this pattern emerged during the Vietnam War, when autopsies were performed on thousands of young soldiers from both sides. American pathologists noticed in the arteries of the U.S. soldiers in their twenties what they called "fatty streaks"—early signs of atherosclerosis. They did not find these streaks in the young Vietnamese, who had grown up in a much less-polluted country. The main reason for the discrepancy was thought to be lifestyle—in particular, diet. But pollution may have been an additional factor, as it promotes inflammation of the arteries.

Examining the large databases of international health organizations such as the WHO, CDC, and the International Heart Association reveals that the rate of CVD appears to track, both historically and geographically, that of pollution from fossil-fuel emissions. Just as global warming has intensified in correlation with the increase in atmospheric carbon dioxide (CO_2), the incidence of cardiovascular disease has risen in correlation with air-pollution levels. And what is driving these trends? The burning of fossil fuels.

After having taken aim at cigarette, soft-drink, and fast-food producers, cardiologists are now turning their sights on oil and coal companies—because examining where and when air pollution is at its worst seems to explain the geotemporal fluctuations in CVD. Air-pollution levels skyrocketed in the developed countries in the first half of the twentieth century, before gradually declining as various regulatory measures took effect and air quality improved. We see a corresponding sharp rise and gradual fall in the rate of CVD deaths in the West. Now that same spike in the incidence of heart attacks and strokes is taking place in the developing countries, where coal and oil are fueling a massive increase in manufacturing output and a corresponding increase in atmospheric pollutants.

This finding throws new light on the alarming statistics on CVD coming out of the nations of the former Soviet Union, where coal has long been the main energy source. The same scenario is unfolding in China, the new global economic marvel, whose factories also run mainly on coal-fired energy. In 2009 an estimated two million people died from pollution in China.[10] Over the past fifty years cardiovascular deaths in China increased by five times; between 1985 and 2005 they more than doubled. These increases parallel that of China's economic growth but do not take into account all of the cars that the Russians, Indians, and Chinese have bought over the past twenty years: approximately half-a-billion cars spew exhaust into the atmosphere in those areas of the world.

In coal- and gas-fired Russia, the CVD death rate is ten times the rate in France, where for several decades energy has been produced in nuclear power plants. It is interesting to note that Russians who have lived in France since birth have the same CVD rates as those of French citizens.

The cardio-environmental model also helps explain another acknowledged disparity: that the CVD death rate in northern Europe, including Scandinavia, has long been higher than that in the Mediterranean region.[11] This disparity, often called the "French paradox," may be related to the Mediterranean diet, as several studies suggest, or to the consumption of red wine, still a somewhat controversial notion. There is, however, another important difference between people who live in the Nordic regions and those who live farther south: northerners must heat their houses in winter and, consequently, are exposed to higher levels of air pollution. Fortunately, the north–south gradient in CVD mortality has eased in recent years. For example, in Finland, which at one time held the world record for heart disease and strokes, CVD deaths have dropped significantly, not only as a result of declines in the classic risk factors for cardiac disease but also from the implementation of antipollution measures.

There is also a west–east gradient in the cardio-environmental model. As westerly winds are prevailing, they push air from the Atlantic coastal region of Europe toward the east, accumulating atmospheric pollutants over the heavily industrialized eastern regions of Europe. And in Western Europe, where antipollution measures have led to marked improvement in air quality, there has been a corresponding decline in CVD rates. However, antipollution regulations are much less rigorous the farther east one goes, as is reflected in the discrepancy in CVD rates between France and Russia. The farther north one goes, the greater the need to heat homes in winter; the farther east, the greater the use of coal-fired energy. The cardio-environmental model seems to hold up.

The Framingham study did not name air pollution as a risk factor for heart disease. This omission is not surprising because, on the one hand, air pollution was not measured at the time and, on the other, the Framingham survey population lived in a relatively homogenous environment, so differences in environment would not have been a factor. Moreover, it was not until 1993 and the publication of the Harvard Six Cities Studies, discussed in the next chapter, that the correlation between air pollution and heart health emerged.

The cardio-environmental model evokes Al Gore's CO_2 curve, which shows the correlation between the rise in atmospheric CO_2 and in the average global temperature. In fact, these several elements seem to be connected: the burning of fossil fuels produces the greenhouse gases and fine particulate matter that drive global warming, ocean acidification, and glacial melting. These same airborne pollutants are driving the rise in arterial disease, heart attack, and stroke.

Global Warming and Inflamed Hearts

The evidence is there, and we cannot claim that it emerged only recently. Once again going back in time, we see that the connection between heart health and air pollution seems to have been known around 400 BC to Hippocrates, widely regarded as the "father of medicine." In his treatise *On Airs, Waters and Places*, Hippocrates suggested that the sole cause of disease was the air people breathe.

During the industrial revolution, particularly in Great Britain, coal fires burned in every home and factory. In 1892 in London, a heavy three-day smog resulted in 1,000 premature deaths. In 1905 the city experienced another "great stinking fog," one outcome of which was the coining of the term "smog" by Henri Antoine Des Voeux for a paper he delivered titled "Smoke and Fog."[12] London's *Daily Graphic* newspaper of June 26, 1905, noted in its coverage of the presentation: "Dr. Des Voeux said it required no science to see that there was something produced in great cities which was not found in the country, and that was smoky fog, or what was known as 'smog.'" The following day the *Globe* newspaper wrote that Dr. Des Voeux had done "a public service in coining a new word for the London fog."

London, December 1952. Extreme cold, heavy emissions from coal- and gas-fired furnaces, and an absence of wind converged to create the conditions for the Great Smog. So dense was the smog that traffic was stopped and people who went outside for a bit of "fresh air" had to wear hospital masks. It lasted just five days but resulted in up to 4,000 excess deaths in one week and 12,000 excess deaths over the year.[13, 14]

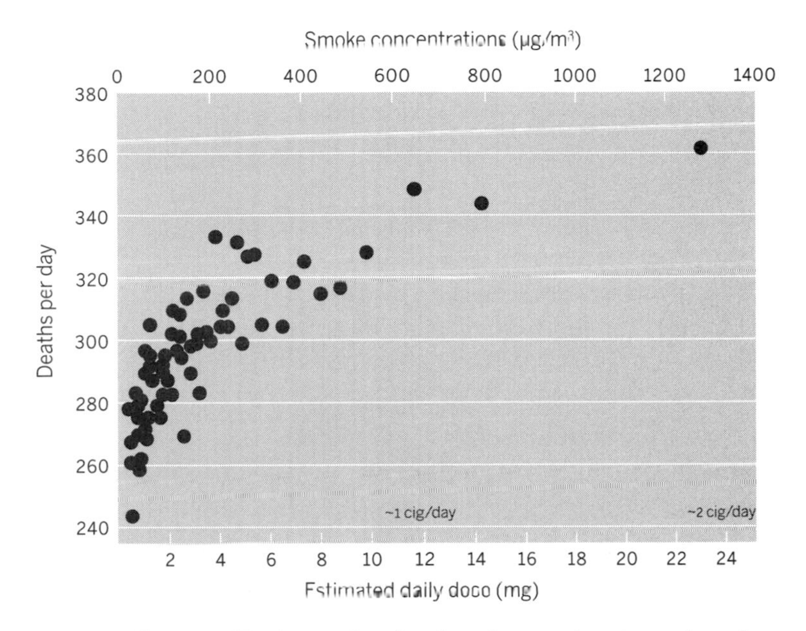

FIG 2 *Daily mortality in London for the winters of 1958–72 plotted over concentrations of particulate air pollution (British smoke, μg/m³) and estimated daily dose.*

Public health authorities in London quickly realized that they had a crisis on their hands. Local hospitals were overwhelmed with the sick, and the number of daily deaths rose from 250 to almost 1,000 at the peak of the episode. Ironically, many in the medical community thought they were dealing with a deadly flu epidemic like the epidemic of 1918. Few in British society were willing to acknowledge that a new horror had descended upon them, a toxic brew spawned by the engines of the nation's economic success—coal and oil. Fewer still were about to stop driving their cars and heating their homes.

The Great Smog did, however, wake up British authorities to the threat of air pollution and set in motion passage of the Clean Air Act. It took four years, and the resultant 1956 legislation regulated smoke emissions mainly by trying to relocate the problem: ordering factories using coal-fired power to move outside the city limits and (after the act was rewritten in 1968) to build high chimneys.[15]

And, in time, British public health statistics began to confirm the premise upon which the act was based. The graph in Figure 2 shows that the daily death rate in London tracks the pollution rate, and it establishes a physiopathological parallel between tobacco smoke, a long-acknowledged pathogen, and industrial and vehicle emissions. Air pollution is toxic to the heart in much the same way that tobacco smoke is.

Twelve Thousand Deaths

By the time of the fiftieth anniversary of the Great Smog in 2002, British health authorities had concluded that 12,000 people had died from this episode of air pollution. The world over, departments of public health have described similar episodes in their cities, and it is indeed revealing to read a chronology of major international air-pollution events and associated fatalities.[16] The history of the industrial revolution is heavily dotted with incidents of massive pollution, among them New York's infamous smog of 1963.

During another environmental crisis, the heat wave in Europe of August 2003, 70,000 people died, including 15,000 in France, where fatalities were concentrated in Paris and the surrounding region. Cardiovascular disease accounted for 72 percent of the deaths in Paris, according to data from the Institut national de la santé et de la recherche médical (INSERM).[17] This figure certainly raised the eyebrows of more than a few cardiologists.

Excessive heat was undoubtedly a factor in the death rates. There were anomalies, however: fatalities were concentrated in urban centers and were much lower in green zones.

The temperatures registered during the European heat wave of 2003 now seem almost normal in many parts of the world. One such place is Arizona, which has become a Klondike, of sorts, for

North American retirees. High-end housing developments are springing up in and around Phoenix, particularly in Scottsdale, as retirees shun Florida—long the first choice for those in their twilight years—to seek the pure air of Arizona. Those who suffer from coronary disease, heart and respiratory ailments, and arthritis simply feel better there.

The average high temperature in summer in southern Arizona is 41 degrees Celsius (106 degrees Fahrenheit), and the average low is 27 degrees (81 degrees Fahrenheit). Europeans experienced similar high temperatures over several days in the heat wave of August 2003, but the high mortality rate in Paris and other large cities was not attributable solely to the temperature. Air quality was a decisive and synergetic factor.

Clean, dry air. This feature is what attracts people to Arizona, though 97 percent of the state's inhabitants live in homes with air-conditioning—a protective factor against the heat; in contrast, 20 percent of Europeans have air-conditioning. Unfortunately, the electricity that runs Arizona's air conditioners is generated mainly by coal-burning power plants. We are no longer talking about a vicious circle but about a downward spiral!

The rise in average global temperatures is being driven by the rise in levels of greenhouse gases from the burning of fossil fuels, and, as we shall see in Chapter 12, this phenomenon is aggravated in urban heat islands. In addition, these same gases are directly toxic to the human heart. Planet Earth is becoming warmer, and our hearts are becoming inflamed.

As yet, no articles have been published on the correlation between smog levels and CVD deaths in the 2003 European heat wave. Despite this dearth of literature, the cardio-environmental model allows us to compare Paris and, for example, the more rural and thus greener Burgundy with respect to the disparities and correlations of temperature, pollution, and green spaces. Death rates were much lower in the *départements* around Paris characterized by more vegetation and less air pollution.[18, 19, 20] The explanation of the differences in local CVD death rates may be linked to the notion of a greater toxicity created when high temperatures, urban heat islands, and air pollution coincide.

There is additional reason for concern. Heat aggravates the harmful effects of pollution by both increasing ozone particulate matter and reducing human resistance to this heat.[21] The higher the temperature, the more ground-level ozone is produced as a result of the photochemical reaction of the polluting emissions. Breathing ground-level ozone (not to be confused with the greenhouse gas called tropospheric ozone) can cause decompensation—the inability of the heart to maintain adequate circulation—in pulmonary and cardiac patients, and even in individuals presumed normal. Over time, these pollutants gradually erode the inner tissues of the lungs and arteries, then one fine smoggy day they fire the fatal bullet.

We have known for centuries that air pollution is detrimental to human health. Every antipollution campaign emphasizes the vital importance of such measures to people with cardiovascular and lung diseases. But exactly how does air pollution promote heart attack and stroke?

ten

A TALE OF SIX CITIES

WE USED TO THINK that the only risk factors for cardiovascular disease were genetics and diet. Few suspected that the air we breathe might be a risk factor too. It's a simple matter of perception. The food we eat and beverages we drink are tangible, while the air we breathe is not, even though on any given day the air we inhale weighs up to ten times the amount of food and liquid we ingest. Let's see... Twelve to twenty-five inhalations per minute, each having a volume of 1 liter (about 1 quart), gives approximately 20,000 liters of air taken into the lungs each day, equivalent to 20 cubic meters. One cubic meter of air weighs 1.2 kilograms (2.6 pounds), so 24 kilograms (52 pounds) of air pass daily through an adult's body, in contrast to 1 kilogram (2.2 pounds) of solid food and 2 liters (2 quarts) of liquid.

Infants breathe, eat, and drink up to three times as much as adults in proportion to their body weight. Young children, therefore, are exposed to much higher concentrations of pollutants than are adults.

It took until the 1970s to prove that smoking is toxic and until the 2000s to learn that exposure to secondhand smoke increases the risk of vascular diseases by 25 percent. This discovery spurred numerous studies on the association between urban air pollution, secondhand smoke, and heart disease. But proving that air pollution causes atherosclerosis, heart attacks, and strokes was no easy task; researchers encountered many obstacles.

Except in cases of extreme smog that can turn the center of a city into a gas chamber overnight, the effects of chronic exposure to air pollution may take decades to become manifest. Researchers had to observe whole cities for years in order to draw conclusions.

Exposure rates vary geographically and are hard to measure. Tobacco exposure is much easier to quantify: researchers only have to count the number of cigarettes smoked in order to calculate the exposure, the risk, and the connection between the two.

Finding funding for research on the association between air pollution and CVD is difficult because the traditional sources for medical research, primarily the pharmaceutical industry, have little financial incentive to invest. Pharmaceutical and other industrial products cannot cure air pollution. The financial burden for this research has therefore fallen on public sources and private sponsors concerned with the environment. Fortunately, many such agencies and patrons have stepped up to the plate—for example the Guzzo family of Montreal, founders of the Guzzo cinema chain and sponsors of the Guzzo Environment-Cancer Research Chair at the University of Montreal.

Finally, to demonstrate the toxicity of fossil-fuel burning is to challenge prevailing notions of industrial progress and invoke resistance from powerful corporate interests. We know how strong this resistance can be because we have seen it at work in the tobacco, soft-drink, and asbestos industries.

Despite all of these obstacles, researchers persisted, and the evidence began to pile up. When in June 2010 the American Heart Association released a major update on the association between air pollution and heart disease, no fewer than 426 scientific studies had been reviewed by leading experts in cardio-environmental health.[1] Here briefly is a review of the findings of some of the most compelling of these studies.

Perhaps the pioneering study on the causes of CVD was the Harvard Six Cities Study (HSCS), published in 1993 in the *New England Journal of Medicine*.[2] The full story of this remarkable study may be found on the website of the Harvard School of Public Health,[3] but let me mention the highlights. Several tenacious Harvard researchers launched their ambitious project in 1973. For sixteen years they followed 8,111 adults in six American cities (Watertown, Massachusetts; Harriman, Tennessee; St. Louis, Missouri; Steubenville, Ohio; Portage, Wisconsin; and Topeka, Kansas), collecting data on a variety of indicators: age, weight, sex, lifestyle (including smoking), and social and medical history. The researchers measured exposure levels to fine airborne particulate matter, acid aerosols, sulfur dioxide, and nitrogen dioxide. They tracked the mortality statistics. The Harvard Six Cities Study was an environmental Framingham.

The authors found that the cardiovascular mortality rate in the survey city with the most pollution was 25 percent higher than the rate in the city with the least pollution, all other factors being equal. The premature deaths from air-pollution exposure were linked to lung cancer in 8 percent of cases and to cardiovascular disease in 55 percent of cases—the first time this link was reported. It is useful to note that the 25 percent difference in mortality was recorded between two American cities in which inhabitants had similar lifestyles. It would be interesting to set up a comparison in which one of the cities had very little or no pollution. The data from such a study would be especially pertinent for cardiology.

Since the Harvard Six Cities Study was first published in 1993, hundreds of corroborating studies have come out, and we now know a great deal about the precise mechanisms and role played by air pollution in cardiovascular health.

Stress: a Heart Stopper

Stress is indeed bad for the heart, but the real culprit is oxidative stress, not the stress that most people think of when they hear the word (a somewhat minor and controversial risk factor, often confused with depression, a known risk factor). Air pollution contributes to the oxidation in the arteries in much the same way that oxidation leads to rust in a metal pipe. In arteries, oxidative stress results in inflammation and atherosclerotic plaque—arterial rust—the fatty substance that interventional cardiologists squeeze aside with stents.

How, physically, does this oxidation occur? When we breathe air containing pollutants released during fossil-fuel combustion, we take in a mixture of gases and very small particles called particulate matter (PM). Airborne particulate matter is categorized according to size: large particulates have a diameter of 10 microns (PM_{10}) or less, fine particulates are 2.5 microns ($PM_{2.5}$) or less, and ultrafine particulates are 0.1 microns ($PM_{0.1}$) or less. To get an idea of the relative size of particulates and other cellular and molecular structures, see Figure 3.

Human arteries are vulnerable to attack by atmospheric hazards precisely because ultrafine particulates are small enough to be carried deep into the lungs and, from there, to make their way into the bloodstream. (The membranes of the lung are finer and more permeable than those of the digestive tract.) Here these fine particulates come into contact with the arterial endothelium, the inner membrane of the arteries, whose role it is to protect and regulate the arterial system. If the endothelium reacts to the particles with inflammation, both short-term and long-term damage may result.

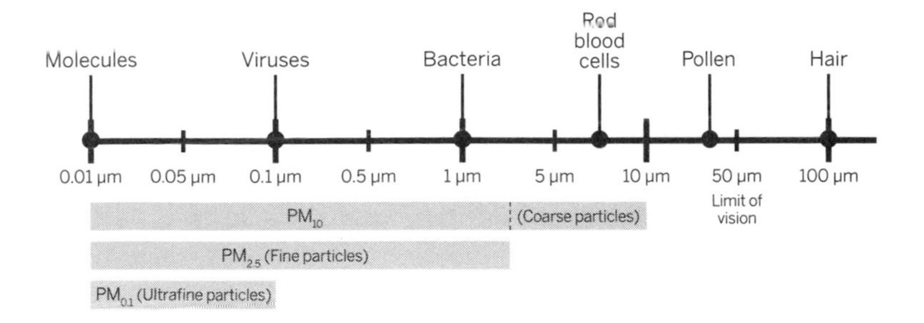

FIG 3 *Sizes of particulate matter in air pollution relative to other common microscopic structures.*

Also entering the body by the same pathway—and causing oxidative stress—are toxic, unstable molecules known as free radicals. In the bloodstream they are usually triggered by such gaseous pollutants as ozone, sulfur dioxide, nitrogen dioxide, and carbon monoxide. Other airborne contaminants that may affect our health include volatile organic compounds (VOCs) and polycyclic aromatic hydrocarbons (PAHs).

In the lungs and arteries, all of these compounds can induce oxidative stress and inflammation, a condition similar to the chemical pneumonitis that afflicts firefighters who inhale smoke from burning buildings. In response, the body produces inflammatory proteins—interleukin, cytokines, C-reactive protein—which in turn damage the inner and mid-level lining of the arteries.[4]

Airborne pollutants also affect other human biological systems by stimulating inflammatory proteins and coagulants, thereby making atherosclerosis plaques vulnerable to rupture and subsequent thrombosis. The autonomic nervous system reacts by constricting the arteries, leading to hypertension and insulin resistance, which in turn may lead to a pre-diabetic condition.

Country Mouse and City Mouse

Valentin Fuster of Columbia University in New York City, a leading authority on atherosclerosis and vascular inflammation, conducted a very interesting laboratory experiment to determine how mice react to air pollution. Lab mice were exposed to two risk factors for CVD: air containing fine particulate matter ($PM_{2.5}$) and a high-fat diet. When the abdominal aortas of the test animals were examined, more atherosclerotic plaque was detected in the mice exposed to polluted air or a high-fat diet or both than in the mice exposed to filtered air and a normal diet.[5]

The dissected aorta of the mice exposed to polluted air, and thus to oxidative stress, shows a larger mass of atherosclerotic plaque than that found in the animals given filtered air. When, in addition, the mice were exposed to the known risk factor of a high-fat diet, an even larger amount of atherosclerotic fatty substance was detected. Here is a laboratory experiment that demonstrates a recipe for a vascular tsunami in humans: eating a diet of fast food in a large polluted city.

Planet Heart, the Cardiologist's Tour

As early as 2006, in a milestone publication in *Circulation Research*, Aruni Bhatnagar had demonstrated that overall mortality in the United States was very tightly correlated with the geographic level of $PM_{2.5}$. As Bhatnagar wrote, "The most surprising aspect of the association between mortality and PM is that pollution is positively and selectively associated with deaths from cardiopulmonary disease. In most studies, statistical associations are stronger for cardiovascular deaths than for all-cause mortality."[6]

Let's take a tour around the world to see what specific links researchers have uncovered between pollution and CVD.

EUROPE: More hospital admissions for pollution-related heart disease

An investigation similar to but even more extensive than the Harvard Six Cities Study was published in Europe in 2002. Looking for evidence of an association between airborne particulate matter and heart disease, researchers examined the records of 38 million hospital admissions from eight European cities. They found that hospital admissions for heart disease rose by 0.5 percent for each observed increment of 10 micrograms per cubic meter (μg/m³) in fine particulate matter and black smoke in the air.[7]

GERMANY: Air pollution's contributions to hardening of the arteries

In the heavily industrialized region of the Ruhr, in western Germany, researchers examined 4,500 residents who lived close to heavy traffic to determine if this exposure represented a risk factor for atherosclerosis and coronary artery calcification (measured by the coronary artery calcification score, which is calculated using CT scans). The conclusion? The closer the participants lived to an area of heavy traffic congestion, the higher their calcification score. Those who lived less than 50 meters (55 yards) from a polluted roadway had 63 percent more coronary calcification than those who lived more than 200 meters (220 yards) away from a high-traffic road—more proof that fine particulate matter released by motor vehicle exhaust is correlated with the level of vascular disease.[8]

The study also filled in a missing link in coronary medicine: Why do some patients have high levels of coronary calcification while in others the condition is absent, even though both groups have similar risk factors? Here was a new factor—environmental risk. To the list of classic risk factors for cardiac arrhythmias, heart failure, thromboses, heart attacks, sudden death, and strokes, cardiologists now had to add exposure to air pollution.

As you can imagine, cardiologists found these and other public health studies to be a revelation.

ITALY: More deep vein thrombosis with chronic exposure to particulates

Researchers in Lombardy, Italy, studied 663 patients suffering from deep vein thrombosis and 859 control subjects between the years 1995 and 2005. They showed that each 10-microgram increment in the chronic intake of PM_{10} increased the risk of deep vein thrombosis by 47 percent.[9]

BOSTON: Increase in heart attacks after a sharp rise in air pollution

Between January 1995 and May 1996, researchers conducted a study of 772 heart attack victims living in the greater Boston area and found that exposure to increased fine particulate matter ($PM_{2.5}$) had apparently triggered acute myocardial infarction (MI) in some of them. A surge of 25 µg/m³ in airborne $PM_{2.5}$ during the final two pre-symptomatic hours was associated with a 48 percent higher risk of MI. A delayed response was also observed. An increase of 20 µg/m³ in atmospheric $PM_{2.5}$ during the twenty-four-hour period preceding symptoms was associated with a 69 percent higher risk of MI. The authors suggest, "elevated concentrations of fine particles in the air may transiently elevate the risk of MIs."[10]

GERMANY: Exposure to heavy traffic strongly associated with subsequent heart attacks

A group of researchers led by Annette Peters examined 700 patients hospitalized for MI in and around Augsburg, Germany, between 1999 and 2001 and reported that traveling in heavy traffic can trigger heart attacks in the hour following exposure, with a risk of MI that is 2.92 times higher than without such an exposure. Thus, the risk of heart attack was nearly three times as great after traveling in heavy traffic.[11]

THE NETHERLANDS: A doubled risk of cardiac death in persons living near a major roadway

A Dutch study that followed 5,000 individuals between 1986 and 1994 showed that those living less than 50 meters (55 yards) from a major roadway were 95 percent more likely to die of a cardiopulmonary event than those who lived 200 meters (220 yards) or more away from any such roadway, all other factors being equal.[12]

FINLAND: Rise in stroke deaths following high air pollution

In Helsinki, an observational study that analyzed the mortality among one million inhabitants for six years (1998–2004) reported a significant association between deaths from stroke (3,265 deaths) and an increase in $PM_{2.5}$ on the day before and the day of death. Stroke deaths rose 6.9 percent for each quartile rise in PM exposure.[13]

BOSTON: Rise in strokes in periods of elevated air pollution

Boston-based researchers found that each time the daily rate of atmospheric pollution increased from "low" to "moderate," the incidence of stroke during the following twenty-four hours rose by 34 percent.[14]

UNITED STATES: Malignant arrhythmias after air-pollution exposure

In a Harvard University study of 203 patients with implanted cardioverter defibrillators (ICDs), researchers reported a significant association between elevated air pollution and the arrhythmias recorded on patients' defibrillators during the three days following exposure. They also found a strong link between fine particulate matter, carbon monoxide, nitrogen dioxide, and ventricular arrhythmias.

The Triggers of Ventricular Arrhythmias (TOVA) study surveyed 1,188 people with cardioverter defibrillators from thirty-one American cities and found that ICD shocks for malignant arrhythmias were twice as likely to occur within one hour of taking a trip in a car. The authors noted that their results replicated those of studies that reported increased incidence of heart attack after exposure to motor vehicle pollutants.[15]

Cardiologists find these last studies to be fascinating, mainly because they reveal a new and unexpected use for a high-tech medical guardian angel: the ICD. This marvelous device protects patients at risk of sudden cardiac death by constantly monitoring heart rhythm and, when an abnormal rhythm is detected, generating an electrical impulse to correct it.[16] The ICD records heartbeats with great precision over several years (the battery life of the device). It has become the cardiologist's black box. Every clinical event reported by a patient can be correlated with data from the defibrillator, down to the second. As it turns out, measurements of this level of precision are valuable to researchers seeking to correlate arrhythmias with exposure to air pollution.

UNITED STATES: Cardiovascular deaths due to increases in air pollution

The clincher among the growing number of studies was a sixteen-year survey of 500,000 people living in 156 American cities. It found that every increase of 10 $\mu g/m^3$ in atmospheric fine particulate matter was associated with a 6 percent rise in cardiopulmonary death.[17] In addition, every quartile rise in fine particulate matter increased the overall cardiovascular risk by 12 percent and the risk of arrhythmia and heart failure by 13 percent.[18] The study's principal investigator, C. Arden Pope, stated that air pollution causes more cardiac deaths than respiratory deaths.[19]

Yet another study covering 9.3 million Americans in 126 cities found that an increase of 1 part per million in atmospheric carbon monoxide (CO) from car exhaust led to a 1 percent increase in all hospitalizations for cardiovascular conditions.[20]

In 2008 the U.S. National Research Council, working under the mandate of the Environmental Protection Agency, declared that "short-term exposure to ambient ozone is likely to contribute to premature deaths."[21] Had the authors used plain language, the statement might have read, "smog probably kills." Frank O'Donnell, president of Clean Air Watch, a Washington-based non-profit devoted to protecting clean-air laws and policies throughout the United States, did not mince words in his follow-up statement. He wrote that "the report is a rebuke of the Bush administration, which has consistently tried to downplay the connection between smog and premature death."[22]

In its 2010 update of the scientific literature, the American Heart Association issued several recommendations on how to reduce exposure to air pollution, which are summarized here.[23]

- Educate all patients with CVD about the dangers of air pollution.
- Encourage patients to consult the available sources for information on forecast air quality.
- Advise patients to reduce exposure and activity during times of high air pollution.
- In the home, when the air quality index indicates high pollution, keep windows closed and use a ventilation system equipped with an air filter.
- Avoid travel to and prolonged stays in areas of high air pollution.
- If possible, avoid travel on heavily used roadways and through areas of heavy industry.
- Maintain optimal car filter systems, drive with windows closed, and recycle the air inside the vehicle.

Reading the list you might well wonder if we were under attack—if environmental terrorists had released a poisonous gas and authorities had invoked emergency measures to stave off mass extermination. At the very least you might conclude that following the old refrain "go outside and get some fresh air" is now a *risky* activity. Note, though, that the air in some interiors, cars included, may be even more polluted than outside air if it is not adequately filtered and ventilated.

How Important Is Air Pollution in Cardiovascular Disease?

Among all of the risk factors for CVD, what is the relative importance of air pollution? There is much debate on this topic. Some authors have estimated that air pollution is responsible for only 2 percent of CVD.[24] They studied the main causes of CVD and cancer and concluded that smoking, diet, and genetic predisposition are the three main—and equally significant—risk factors.

Yet the Harvard Six Cities Study reported a 25 percent difference in cardiovascular *mortality* between two of the six survey cities—namely, the city with the highest level of pollution and the one with the lowest pollution—and there's a big gap between this figure and the 2 percent of CVD estimated above. In addition, the data for the HSCS were collected in American cities; no comparisons with non-American cities were undertaken. The eminent environmental economist C. Arden Pope believes that urban air pollution is the thirteenth-leading cause of death internationally. Based on WHO data, he estimates it to be directly responsible for more than 700,000 cardiovascular deaths each year.[25] Moreover, a 2013 survey by the WHO estimated the number of excess deaths from air pollution at more than seven million.[26]

Establishing a consensus on the health impact of air pollution is difficult because, although air pollution is ubiquitous, exposure rates vary, and establishing valid comparisons is problematic, even between two residents of the same area. But the Harvard Six Cities Studies as well as the hundreds of scientific papers that came after it show a clear link between air pollution and cardiovascular mortality.

It is also hard to draw conclusions on the historical evolution of heart disease. Before the 1970s, data collection on environmental risk factors was limited, and clinical databases were smaller and less precise than those available to scientists today. The data collected in the United States after the Second World War and in China in recent years seem to show, however, that the impact of air pollution on CVD is much greater than the 2 percent cited above.

According to a 2006 WHO report, an estimated 24 percent of the global disease burden and 23 percent of all deaths can be attributed to environmental factors.[27] T.J. Grahame and R.B Schlesinger, in a major analysis published in 2009, concluded that emissions from motor vehicles represent a major environmental cause of morbidity and cardiovascular deaths in the United States.[28] David R. Boyd and Stephen J. Genuis estimated that environmental factors account for between 7.5 percent and 23 percent of all heart disease in Canada.[29]

In the twenty first century it is rare to find an unpolluted place to conduct an exhaustive comparative study. But that's the coup that Michael Gurven pulled off with his study on the Tsimane people (see Chapter 2). Others have followed his example and studied aboriginal peoples living in areas untouched by environmental contaminants. These researchers have also reported very little vascular disease and hypertension among their cohorts who, from a Western perspective, are valuable control groups or, rare indeed, human placebo groups, because their environment is almost completely free of pollution.

Air Pollution and the Classic Risk Factors

Not only is there growing proof of a direct causal relationship between atmospheric pollution and heart disease; there is now evidence that pollution aggravates the classic risk factors for CVD that were first identified in the 1980s. In fact, we now realize that the rates of hypertension, metabolic syndrome, diabetes, and even obesity are higher in areas of high atmospheric pollution.

Air pollution as a cause of hypertension

Not so long ago medical students were taught that hypertension was "essential"—had no known cause—in over 90 percent of cases, namely that genetic predisposition was the sole cause. Recent studies have shown, however, that atmospheric pollutants cause arteries to narrow and blood pressure to rise.[30, 31, 32] In other words, arterial hypertension is not unavoidable or innate but, instead, widely environmentally induced. The more pollutants in the air, the more hypertension.

Lead exposure as a cause of hypertension and heart attacks

Some of the health impacts of lead exposure were discussed in Chapter 7. A potent environmental hazard, lead is present in our environment mainly as a by-product of the burning of fossil fuels, particularly leaded gasoline. A joint 2007 study by Harvard University and the University of Atlanta toxicology department tested 837 men for lead levels in either blood or bone (bone provides the best long-term indicator of lead exposure) and followed them between 1991 and 2001. The authors observed 73 percent more angina and heart attacks in those whose lead level was greater than 5 micrograms per deciliter (µg/dL) than in those with less than 5 µg/dL. When they corrected for all risk factors, one standard deviation increase in blood lead level was associated with a 27 percent greater risk of ischemic heart disease.[33]

The ban on leaded fuel for road vehicles in, for example, 1990 in Canada and the United States and in 2000 in China and several other Asian countries has had major health benefits globally. These prohibitions represent a true environmental "prescription" for Planet Heart.

Air pollution as a promoter of diabetes and obesity

Even more intriguing was the discovery that air pollution can cause diabetes. In one study researchers demonstrated that fine particulate matter activates the autonomic nervous system and disturbs insulin secretion by the pancreas. This particulate matter also induces insulin resistance as a result of systemic inflammation and oxidative stress.[34] Another U.S. study showed that industrial pollutants are linked to diabetes.[35]

Researchers from Columbia University in New York reported, astonishingly, that infants whose mothers were exposed to high levels of polycyclic aromatic hydrocarbons (PAHS) during pregnancy had a 79 percent higher rate of obesity at age five years. Polycyclic aromatic hydrocarbons in the environment come mainly from car exhaust. This study provided evidence of a significant *in utero* hazard and environmental cause of obesity: a pregnant mother exposed to air pollution.[36]

It has also been shown that exposure to polychlorinated biphenyls (PCBS) is linked to the growing prevalence of diabetes. Between 1929 and 1971, the city of Anniston, Alabama, was a major manufacturing center for PCBS, and residents of the city had some of the highest levels of PCB exposure ever recorded. One study of residents under the age of fifty-five found that those with the highest exposure levels (the highest quintile of blood-level PCBS) had a rate of diabetes 4.78 times the rate of those in the lowest quintile.[37]

In addition, we have learned that certain pesticides lead to insulin resistance and pre-diabetes in mice. The researchers who made this discovery went on to suggest that pesticides in our environment, including even those from poorly washed vegetables, may be linked to the sharp rise in diabetes rates around the world.[38] Such correlations have yet to be demonstrated in humans, however.

Air pollution and diabetic mortality

During periods of extreme air pollution, persons with diabetes are at two times the risk of heart attack or cardiovascular death of those without the disorder. The oxidative stress from air pollution exposure exacerbates existing damage to the arteries caused by diabetes.[39, 40, 41]

Moreover, metabolic syndrome, dreaded by patients and doctors alike for its damage to blood vessels, is also associated with atmospheric pollution. The term "metabolic syndrome" is something of a medical catch-all used to describe the complications related to excess weight, such as diabetes, hypertension, and abnormal cholesterol or triglyceride levels that put people at higher risk of heart attack. In fact, it properly refers to a constellation of metabolic problems linked specifically to abdominal obesity. Air pollution seems to accentuate and potentiate the harmful effects of cardiometabolic syndrome on the arteries.[42]

Air pollution and dementia

Vascular dementia is caused by a series of small strokes in the brain. Although these "mini-strokes" often go undetected, they lead inexorably to severe loss of higher cognitive functioning. Dementia seriously compromises patients' quality of life. And dementia rates are skyrocketing in the developed countries, mainly owing to the aging of these populations. Vascular dementia is the second most common form of dementia after Alzheimer's disease. However, vascular dementia and vascular disease itself are preventable, unlike Alzheimer's disease, for which no effective protective treatments are known.

Researchers have recently discovered that environmental factors play a major role in the growing rate of age-related dementia. A large American study funded by the National Institute of Environmental Health Sciences, the National Cancer Institute, and the Environmental Protection Agency found a link between exposure to high levels of air pollution and a rise in the incidence of dementia. The authors discovered that living in a polluted area accelerated mental aging by an average of two years, mainly as a result of the harmful effects of fine particulate matter on the arteries of the brain.[43] In addition, researchers in Rochester, New York, found that high levels of lead in blood and bone are associated with more cognitive deficits and the development of dementia.[44] Recall the silent killers of the ancient Romans. Well, we have come full circle.

ADDING IT ALL up, we now know that air pollution is linked to hypertension, obesity, diabetes, and cardiometabolic syndrome—all of which are among the causes of cardiovascular disease—as well as to dementia. Moreover, atmospheric pollutants can directly damage arteries, meaning that the process is both indirect and direct. Toxicity to artery walls from atmospheric pollution must now be added to the classic risk factors for heart disease.

Given that life expectancy continues to rise, do we really need to worry about air pollution? In the developed world, after all, living to the age of eighty and beyond is no longer the exception, despite the pollutants in the air we breathe. For at least a partial answer to this question, let's look at some statistics.

In North America cardiovascular disease exacts a heavy toll. Each year in Canada there are approximately 450,000 hospital admissions for and 73,000 deaths from CVD.[45] David R. Boyd, a recent Trudeau Scholar at the University of British Columbia and a leading environmental lawyer and author, has estimated that in Canada air pollution is responsible for between 5,000 and 11,000 cardiovascular deaths annually (of 10,000 to 25,000 environmentally related deaths from all causes) and 33,000 to 67,000 hospital admissions for cardiovascular disease (of 78,000 to 194,000 environmentally related hospitalizations from all causes). The corresponding economic and social costs from all environmentally related causes in Canada include $9.1 billion and 1.5 million days spent in hospital every year.[46] Cardiac disease represents half of all environmentally related diseases in this country, and, to the best of our knowledge, air pollution is responsible for up to 25 percent of cardiovascular disease.

Recent reports from China and the World Health Organization confirm these trends. In 2007 the government of China estimated that one million premature deaths annually in China are attributable to pollution.[47] In 2011 the WHO analyzed a vast amount of data collected in 1,100 cities in ninety countries and concluded that two million deaths annually are directly related to pollution.[48]

As we learned in Chapter 1, the WHO has estimated that by the year 2030 some 23.6 million people will die annually from any cardiovascular disease, mainly heart disease and stroke. The OECD Environmental Outlook Baseline scenario has projected that urban air quality will continue to deteriorate globally. By 2050, outdoor air pollution is projected to become the top cause of environmentally related deaths worldwide. At the Climate and Clean Air Coalition conference in April 2013 in Paris, Maria Neira, director of public health and environment for the WHO, estimated that the annual rate of excess deaths from air pollution now exceeds seven million. The more scientists measure this association, the more they find out.[49]

But as all patients explain—and everyone, everywhere, says loud and clear—the issue is not life *expectancy* but *quality of life*. Medical science is very effective at keeping people alive, but chronic illnesses have a high morbidity and cardiovascular diseases are among the most debilitating. The WHO also predicts that the greatest challenge to human health during the next twenty years (2010 to 2030) will be non-transmissible diseases, especially diseases linked to the environment, to genetics, and to the interaction of these factors.[50] Age-old scourges such as smallpox and polio have all but disappeared, only to be replaced by vascular disease and cancer.

The fundamental question now is this: Can antipollution measures reduce cardiovascular disease and improve heart health?

eleven

THE WEST WIND

IT *IS* POSSIBLE to reduce the number of heart attacks by improving both interior and exterior air quality. A number of studies have proven that. But trying to accomplish this goal is an enormous and ever-expanding challenge because global economic growth is currently reliant on the massive consumption of fossil fuels. And although many developed countries have measures in place to protect air quality, the combustion of fossil fuels is rising exponentially, largely because of galloping economic growth in the developing countries, where pollution is also rising exponentially. These countries—in particular the two Asian economic powerhouses, China and India—have been slow to adopt clean-air legislation; they consider the cost of improving air quality to be too high and, therefore, have not made it an economic priority.

One deeply ironic consequence of this global shift: Los Angeles, which now boasts some of the strictest antipollution regulations anywhere in the world, often finds itself blanketed by thick smog carried across the Pacific Ocean from Asia by the prevailing winds from the west. To use a sailing metaphor, in a regatta you want to be *up*wind rather than *down*wind, and in the environmental regatta, Los Angeles finds itself downwind to Asia.

The same often holds true within cities. Upscale neighborhoods are typically in the west end, for the simple reason that the prevailing winds blow from a westerly direction, in accordance with Earth's rotation. Neighborhoods in a city's western section are normally not exposed to the pollution picked up by air currents passing over the downtown and industrial areas, which are situated to the east. Montreal is a perfect example of this arrangement: its wealthiest conclaves are in the western suburbs, its middle- and working-class communities are close to the city center, and its heavy industry is located mainly in the east end. It's the same in Paris, where the chic suburbs of Neuilly-sur-Seine and the Seizième Arrondissement hug the city's western flank. In addition, upscale neighborhoods are typically canopied with trees, while lower-income areas feature barren stretches of cement and asphalt. In urban landscapes, it seems, clean air and trees are reserved for the rich.

Quebec is in a similar situation with respect to its neighbors to the west and south. Located in the easternmost region of the Saint Lawrence river basin, Quebec receives atmospheric pollutants from Ontario and the midwestern United States carried by water (the Saint Lawrence River is the drainage basin for the Great Lakes) and air (the prevailing winds follow the river's route). These pollutants carry potent health hazards.

Interior Air and Heart Attacks

We know from numerous studies, in particular the compelling InterHeart Study,[1] that secondhand smoke increases the risk of heart disease by up to 25 percent. In the early 2000s, concerned about the emerging data, local governments in North America and

Europe began adopting regulations to restrict cigarette smoking indoors in public places. The regulations met with little opposition—quite the opposite—notably because the number of people who smoked had fallen off sharply. Such a policy would be harder to enact in China, where 60 percent of men are smokers, in contrast to 3 percent of women. The percentage of smokers in China has changed little since before the country's industrial revolution, and unfortunately smoking remains a widely accepted lifestyle choice.

In places where bans on smoking in public places were adopted, health authorities observed a positive impact on cardiovascular health—that is, declines in the number of hospital admissions for heart attack. The incidence of heart attack fell, for example, in Rome (by 11 percent), Saskatoon (13 percent), Montana (16 percent), Scotland (17 percent), Colorado (27 percent), and Ohio (39 percent). Interestingly, the lower heart attack rate was reported for both smokers and non-smokers. Here was evidence that the regulations on secondhand smoke were justified.[2]

Exterior Air and Heart Attack

Using the policies on secondhand smoke as a model, we could posit that if exterior air pollution contributes to the onset of CVD, reducing exterior air pollution would lead to a decline in death rates and medical consultations for CVD. We have one wonderful example of how this would play out: the Olympic Games, the epic symbol of human health.

The Atlanta Games

The 1996 Summer Olympics in Atlanta, Georgia, were a showcase for more than athletic prowess and beauty. Although perhaps unintentionally, they also demonstrated to the world how the health of the environment could be improved. Before the Atlanta Games began, city officials passed a number of regulations aimed at restricting automobile traffic in downtown Atlanta, and indeed over the seventeen days of the Games rush hour traffic declined 23 percent. Consequently, the level of ground-level ozone fell by 28 percent and emergency room visits by children suffering from asthma attacks dropped by 42 percent.[3]

The Games in Beijing

Health statistics on China tell us a great deal about the vast changes to the country since its economic boom. Fifty years ago, heart disease, stroke, and cancer accounted for only 16 percent of deaths in persons in China forty years of age and over. Today, these diseases account for 66 percent of all such deaths[4]—a fourfold increase in just two generations. China's current situation parallels that of the United States in the first half of the twentieth century, when cardiovascular disease was rampant. The incidence of stroke deaths has skyrocketed in China as well; in 2008, it was about seven times the rate in the United States.

For the 2008 Summer Olympic Games in Beijing, Chinese authorities put in place strict air-pollution control measures for Beijing and the surrounding area. Vehicle use was controlled and factories were closed in Beijing itself, and coal-burning factories were shut down in nearby Shanxi province. Authorities wanted to prevent not only the smog and carbon pollution that would limit visibility and create traffic turmoil but also, of course, the poor air quality that would affect athletic performance.

A study published in 2009 documented the results of the air-quality control measures undertaken in and around Beijing that summer. Compared with the same period the year before, the level of emissions during the Games was sharply lower. Black carbon, carbon monoxide, and ultrafine particles from fossil-fuel combustion fell 33 percent, 47 percent, and 78 percent, respectively. As dramatic as these results were, there has been no follow-up study on the morbidity impact of the controls. Researchers did, however, seize the unique opportunity of the Games to measure biomarkers of vascular inflammation and thrombosis in 125 young healthy subjects. They observed that the concentration of various particulate and gaseous pollutants decreased substantially (by 13 to 60 percent) from the pre-Games period to the two weeks of the Games. In conjunction, they observed in these healthy subjects significant improvements in inflammatory vascular biomarkers such as sCD62P (P-selectin) and von Willebrand factor. After the Games the rise to "normal" levels of atmospheric pollutants was associated with increases in fibrinogen, von Willebrand factor, heart rate, sCD62P, and sCD40L, all of which are linked to an increased risk of cardiac events.[5]

Pollution, Cardiac Death, and Life Expectancy

After the sale of coal was banned in Dublin, Ireland, in 1999, the city's cardiac death rate fell by about 10 percent within a few years. The ban on coal sales led to a decrease in average black smoke concentrations of 36.5 $\mu g/m^3$ (70 percent).[6] Unfortunately, the coal ban failed to substantially improve air quality in Dublin because numerous residents replaced coal with other particulate-emitting fossil heating fuels. The city would have been further ahead had it followed the lead of Zermatt, Switzerland, and become a combustion engine car-free zone. Zermatt adopted this measure in order to control the air pollution that threatened to obscure the view of the Matterhorn, the town's main tourist attraction.

In 2009, C. Arden Pope, one of the world's leading researchers in environmental economics, found that a decrease in air-pollution levels in American cities was associated with a decrease in mortality and an increase in life expectancy. His study, published in the *New England Journal of Medicine*, was impressive for both its findings and its scope.[7] Between 1980 and 2000, researchers collected data on 116 communities in fifty-one cities, representing a total population of four million people.

Perhaps its most startling conclusion was that a reduction of 10 $\mu g/m^3$ of fine particulate matter air pollution ($PM_{2.5}$) was associated with an overall increase in life expectancy of 0.6 years. In some communities in the period 1997 to 2001, increases in life expectancy of up to five years were observed. This outcome is remarkable, far surpassing the effectiveness (in additional years of life) of the most up-to-date medical technologies and treatments (defibrillators, stents, and, for example, statins), or any gene therapy.

The data for this study were collected during an important time in the history of U.S. air-pollution standards and regulations, a time of intermittent advances and setbacks that today seem symbolic of the eternal conflict between politics and science. In 1979, under President Jimmy Carter, the U.S. Environmental Protection Agency drew up a national plan to restrict emissions of particulate matter sized PM_{15} and $PM_{2.5}$. In 1983, under Ronald Reagan, the government abolished the program. It was not until 1997 during Bill Clinton's term that emission standards were once again put in place in the United States with the adoption of the National Ambient Air Quality Standards.

A Walk in Hyde Park

Within any city the level of air pollution varies from place to place, and so does its resulting health impact. This relationship was shown by a simple but elegant study from London, England, that asked sixty adults suffering from mild or moderate asthma to walk for two hours along a city street (Oxford Street) and, on another occasion, two hours in a park (Hyde Park).[8] During and after these walks, researchers measured ambient air pollution and the participants' respiratory function. They found that on the street, as compared with the park, the level of ultrafine particulate matter was 63.7 versus 18.3 (x $10^3/cm^3$) and the level of nitrogen dioxide was 142 versus 21.7 ($\mu g/m^3$). The researchers detected a significant decline in respiratory capacity when participants walked in the street, where they had significantly higher exposures to $PM_{2.5}$, ultrafine particles, elemental carbon, and nitrogen dioxide than in Hyde Park. The study demonstrated that our choice of route when we walk or jog—even a difference of a few hundred meters—makes a difference to the activity's health impact.

Dark Clouds on Health's Horizon

Air pollution is carried across borders within continents and also across oceans from other continents. An estimated 25 percent of Los Angeles's air pollution originates thousands of miles away in Asia and the Middle East. This pollution is generated in megacities such as Bangkok, Beijing, Cairo, Dhaka, Karachi, Calcutta, Lagos, Mumbai, New Delhi, Seoul, Shanghai, Shenzhen, and Teheran. If heavy rains, such as the monsoons, do not wash the aerosol particles from the air, atmospheric brown clouds appear.

Known as ABCs (the term was coined by the United Nations Environmental Programme), these clouds form when particles such as soot and black carbon from the burning of fossil fuels and biomass exceed 10 percent of the particulate matter present in the atmosphere.[3] These brown clouds can be up to 3 kilometers (1.9 miles) thick and cover vast areas.

ABCs are also found elsewhere, notably over the Amazon basin, where they are generated by the slash-and-burn deforestation (biomass combustion) widely practiced in the Amazon rain forest.

Their tons of aerosol pollutants can decrease by up to 25 percent the amount of sunshine hitting Earth's surface. ABCs may be contributing to climate change and affecting weather patterns, glaciers, and rivers. They rise into the troposphere, reaching up to 5,000 meters (16,400 feet) in altitude and, as NASA satellite images show, transport pollution from continent to continent in an eastward direction on the prevailing winds. On the receiving end are the west coasts of Canada and the United States. In certain parts of California, for example, up to 40 percent of the air pollution originates in Asia.

Pollution is globalizing, and so too is coronary disease.

twelve

DOCTORS OF PLANET EARTH

WHO IS THE DOCTOR in the house when it comes to biotopes or habitats, you might be wondering? Well, among the professionals devoted to maintaining human health and well-being, some are generalists and others are specialists. The same is true for the doctors of Planet Earth; the generalists—environmentalists—work in teams with specialists such as geographers, cartographers, and biologists.

Guy Garand, director of the Regional Environmental Council in Laval, Quebec, was a generalist who took this approach of teamwork, developing and leading a study called Projet Biotopes in collaboration with the Montreal Metropolitan Community (MMC) and specialists (two geographers, one botanist, and one epidemiologist) from the geography departments of the Université du Québec à Montréal (UQAM) and the University of Montreal, as well as the latter's Institut de recherche en biologie végétale (Plant Biology Research Institute). This five-member team—environmentalist Guy Garand, geographers François Cavayas and Yves Baudouin, botanist Yann Vergriete, and epidemiologist Norman King—assembled to examine the phenomenon of urban heat islands (UHIs) within the MMC. The researchers, pioneers in Quebec in the use of geospatial technology, concluded incontrovertibly that the incidence of UHIs is directly related to urban deforestation.[1]

Using images taken by the Landsat 5 satellite, which has been photographing Canada from space for almost thirty years, the researchers analyzed and compared images of the MMC region taken in 1985 and 2005 to discover the nature and location of the region's urban heat islands. They also gathered quantitative soil data—in particular, data on soil temperatures as measured by thermography.

Urban heat islands are zones in a city where average temperatures are significantly higher than those in neighboring zones. In summer, temperatures in such hotspots may be from 10 to 15 degrees Celsius (18 to 27 degrees Fahrenheit) higher than temperatures in nearby areas, and differences of 20 degrees Celsius (36 degrees Fahrenheit) are not unheard of. The phenomenon was first observed and named by meteorologists, though it has long been familiar to, for example, glider pilots who, while soaring and passing over a village or a large expanse of concrete or pavement, may encounter what is called a "thermal" (a column of rising air formed because of ground heat) that can hold the glider aloft indefinitely, or at least until nightfall. Because of such hotspots, cities create microclimates, and megacities engender changes in climate that are measurable on a meteorological scale.

A satellite image of Saint Laurent, a suburb of Montreal, taken one fine June day in 2005, revealed that it was 23 degrees Celsius in the wooded area, 27 on the golf course, 31 in a residential area with light vegetation, and 40 degrees in a heavily paved and built industrial zone. The distance between the zone with the lowest recorded ground temperature and the one with the highest was less than 500 meters, yet the temperature difference was 17 degrees.

A difference in air temperature of even 10 to 15 degrees is enormous in terms of human health, and heat can be a serious health threat, especially for Westerners whose bodies have been "reset" and are losing the ability to tolerate the rigors of climate. The Tuareg and Maasai peoples smile sardonically when they see tourists arrive from the West, carrying huge amounts of water because, back home, the thermostats of their houses, offices, and cars are set at 21 or 23 degrees Celsius (70 or 73 degrees Fahrenheit) and they, like most Westerners, have become extremely sensitive to and also vulnerable to climatic variations. Heat stroke can be fatal for frail elderly persons and those suffering from chronic diseases—cardiovascular disease, in particular—as well as diabetes and pulmonary disease. Heat stroke can also affect people in good health, if they have lost the ability to adapt to temperature variations.

The condition of having a normal body temperature is known as normothermia. Scientists have shown that excess normothermia contributes to obesity, because the human body is more likely to maintain its ideal weight when it must react to temperature variations, especially to harsh climatic conditions. People burn energy, and thus fat, to warm themselves and also to cool themselves. If, however, they are exposed to a reliable, constant supply of heat or air conditioning, they will burn less energy and probably gain weight. North Americans rarely go outside nowadays; barely 5 percent of daily activities take place outdoors. Modern buildings and transportation services in cities make it possible to live a normal life without ever going outside!

As research in kinesiology has shown, people burn more calories when engaged in an outdoor physical activity than when doing the same one indoors because, outdoors, the body must heat or cool itself, depending on the situation. The product of many thousands of years of evolution, the human body has become a magnificent physiological mechanism, but it is atrophying. The average North American expends almost no calories to maintain his or her body temperature. Here, then, is a new perspective on the physiological benefits of the age-old parental refrain, "Go outside and play!"

As we saw in Chapter 9, heat aggravates the harmful health effects of pollutants. It increases ground-level ozone, including the oxidative radicals that harm lungs and arteries, as a result of a thermochemical reaction. One study showed that in the cities of Atlanta and New York a rise in exterior temperature from 10 to 30 degrees Celsius (10 to 86 degrees Fahrenheit) saw a corresponding sixfold increase in ground-level ozone, which is produced mainly by fossil fuels.[2]

Another startling discovery: in Helsinki, the incidence of fatal strokes rose in conjunction with an increase in air pollution, but—stranger still—the effect was significant in summer and not in winter, at least as reported for that Nordic metropolis. A variety of possible explanations may account for the difference, including that people spend more time outdoors in summer and therefore breathe more of the dirty exterior air that can cause strokes, and also that, as was observed in Atlanta and New York City, heat aggravates the harmful effects of airborne pollutants. High temperatures make it easier for fine particulates to penetrate the membranes of the lungs, thus increasing their toxicity.

Because there seems to be evidence that heat increases air pollution, we must determine the causes of urban heat islands in order to find ways to alleviate them. Taking this action will in turn help prevent the excess deaths associated with heat waves and exacerbated by urban heat islands.

When it compared images and data for the MMC in 1985 and in 2005, the Projet Biotopes team came up with such stark and categorical evidence that Yves Baudouin, head of the geography department at UQAM, asked the doctoral students working with him to redo their data analysis a second and third time. The team found very strong evidence that, among all the potential causes of urban heat islands that they had identified, deforestation was the main culprit. Simply put, if you chop down a tree, the ground temperature increases. If you clear-cut a forest, an urban heat island is created. The team observed that UHIs had formed in precisely those areas of Greater Montreal where trees had been cut to make way for commercial and residential real-estate developments and highways.

In their paper on the findings of Projet Biotopes, François Cavayas and Yves Baudouin were able to confirm that the MMC's most excessive heat from UHIs occurred in downtown Montreal. This area was also the portion of the MMC with the highest level of air pollution based on the air quality index (AQI) used in Montreal, a scale from 1 to 50 that indicates the overall level of major air pollutants. In downtown Montreal in 2010, only one in three days was categorized as a "good" air quality day, with an AQI between 0 and 25; this region had the worst air-pollution ratings in Quebec.[3]

What is the effect on cardiovascular health when air quality is "moderate" (AQI of 26 to 50) or "poor" (AQI greater than 50) two out of three days a week? In a recent study in Boston, a city similar in many ways to Montreal, researchers observed a 37 percent increase in strokes during days with "moderate" air quality as compared with "good" air quality days.[4]

Air-quality reports are notoriously dry, which may be one reason so few people consult them. They can also be somewhat confusing, notably because the scales for reporting air pollution are often different depending on the reporting agency—municipal, provincial/state, federal, or international. Some use metric measurements, such as micrograms per cubic meter (μg/m^3); others use the older ppb (parts per billion) designation. In addition, the dosage and exposure scales may vary from one agency to another.

Air quality is often expressed as an index number calculated by combining different ratings, and it can vary significantly. Air quality is sometimes represented by the AQI and sometimes by the air quality health index (AQHI), which indicates the health risk, with a ranking of 1 to 3 indicating low risk and one of more than 10 indicating a high risk.

After consulting the various scales, you might find that the same level of air pollution would be rated "good" in Calgary, "moderate" in Quebec City, and "poor" in Geneva. (This discrepancy would be analogous to the same person's being considered diabetic in Zurich but not in Ottawa, hypertensive in Porto but not in Washington, obese in Paris but not in New Orleans.) These scales are arbitrary because, as we now know, there are no safe or "normal" limits for atmospheric pollutants. They should simply not be in the air we breathe, nor in our blood, and, as with lead, the exposure limits are constantly being lowered.

Moreover, some countries do not measure certain indices. France, for example does not measure pollutants to the size of $PM_{2.5}$ but only to PM_{10}. In China, the only official air-quality reports are issued by the U.S. embassy.

The regional disparities in the definitions of diabetes, hypertension, and dyslipidemia were for many years problematic for clinical medicine, until international committees met to hammer out standardized definitions of these conditions. Environmental medicine is in need of a similar house cleaning to improve understanding and diagnosis of the parameters, the basis of all effective action.

Classical medicine has become incredibly precise. We can measure even minute quantities of lipoproteins, inflammatory vascular biomarkers, the proteins expressed by our genome, and measure in the order of a few tenths of a micron the dimension of an atherosclerotic plaque. But who among us knows the extent to which we have been exposed to and our bodies have absorbed PM, VOCS, PCBS, and HAPS? Twenty-first-century medicine is extremely interested in what goes on *inside* us but much less interested in what is going on *around* us.

Given our knowledge of the physiological effects of air pollution, even thinking about the high air-pollution levels in Montreal in the 1970s is enough to make a person shudder (and the situation was worse in the preceding decades). Fortunately, the current lower levels of atmospheric pollutants may be responsible, at least in part, for today's lower rates of heart disease.

This cardio-environmental good luck may be in peril, however. There has been little progress—some would say none at all—in reducing the concentration of atmospheric fine particulates ($PM_{2.5}$) in Montreal, and scientists are predicting that this environmental hazard will worsen in coming years in conjunction with regional expansion.

What's more, the Projet Biotopes team came to a disturbing conclusion after analyzing satellite images of the MMC: the rate of deforestation was still rising on the Island of Montreal. "Since the 1960s there has been a steady decline in woodlands in designated urban zones. Whereas in 1965 the island was 25 percent wooded, by 1995 it was less than 15 percent wooded, and woodland is now declining by about 7 square kilometers per year," he wrote, noting that 18 percent of the island's woodlands disappeared between 1998 and 2005. "It's time to act to save our forests from destruction because land speculators have their sights on them."[5]

Projet Biotopes made excellent use of geospatial technology to diagnose a situation that threatens the health of the environment in and around Montreal. The team also proposed a solution, which they backed up with evidence: stop cutting trees because vegetation mitigates urban heat.

The Sleeping Volcano

The drastic loss of urban greenery can only amplify the threat of urban heat islands, which are, in many respects, urban volcanoes in the making. Heat islands are significant sources of ground-level ozone and can also aggravate the hazards of other pollutants. Perhaps the most shocking example of this effect was the heat wave of

August 2003 that killed 70,000 Europeans, most of whom lived in urban centers where vegetation had become scarce.

Erich Fischer and Christoph Schär, Swiss climatologists from the École polytechnique fédérale de Lausanne, reported in a 2010 article that the average number of heat waves in Paris between 1961 and 1990 was 1.5 per year. They projected that between 2021 and 2050 the number of heat waves will average 13 per year, before climbing to 20.5 per year between 2071 and 2100. In addition, they projected that the Mediterranean region will have 41 days of heat wave per year, or ten times the current rate.[6] As worrisome as these figures are, other climatologists say that this study underestimated the potential health risk of heat waves because the models on which it was based did not factor in the effects of urban heat islands, even though UHIs may aggravate local temperature increases.

A similar study that would factor in the effects of urban heat islands is needed for North America. When it comes to environmental cardiology, we still have much to learn.

Nonetheless, the phenomenon of UHIs and their adverse health effects is widely acknowledged internationally, and numerous cities are reversing the damage of deforestation by intensively planting trees to help protect their residents from the consequences of both heat waves and rising levels of atmospheric pollution. Trees offer numerous benefits: they help to cool neighborhoods, stabilize soil and embankments, protect against floods and sand storms, and capture carbon dioxide and particulate pollution. Beijing, Los Angeles, New York, and Paris have massive vegetation restoration programs and have planted millions of trees to restore the balance between the natural environment and residents.

Journalist Louis-Gilles Francoeur, who has covered the environment for thirty years, once quipped when we both appeared live on Radio Canada's "L'après-midi porte conseil": "It's great to see that the more trees and birds there are in an urban environment, the more human the environment is."

The Collapse

History abounds with mysterious cases of lost civilizations: Easter Island and the Maya cities come immediately to mind. The horrifying notion that a city or even an entire civilization could simply disappear was imagined in the legend of Atlantis long before humans began to excavate actual examples of such massive losses and to identify the factors that may have caused them. This subject is one that Jared Diamond has thought deeply about.

After graduating from Harvard University in 1958, Jared Diamond completed his doctorate in physiology at Cambridge University in 1961. He was appointed a professor of physiology at the University of California, Los Angeles (UCLA) Medical School in 1966. Diamond then embarked on a second career as a biologist, studying ecology and the evolution of the birds of New Guinea. In the late 1980s his interests expanded to include the history of the environment, and he later joined UCLA's geography department as a professor, a position he still holds. Diamond is the author of numerous scientific publications and was awarded the National Medal of Science in 1999. Among his best-selling popular science books are *Guns, Germs, and Steel*, for which he won a Pulitzer Prize in 1998, and *Collapse* (2005).

I trace Jared Diamond's career path simply to show that his interest in animal physiology has evolved into an interest in the "physiology" of the environment. From humans, to birds, to his own environment, this celebrated thinker now seeks to understand humanity in relationship to its planetary home.

Examining in *Collapse* the disappearance of a dozen or so legendary cities and societies from different eras, Diamond argues convincingly that several common factors came together to create the conditions for their severe degradation or destruction. Foremost among those factors was deforestation—local inhabitants cutting down the trees surrounding their settlements. Then, for lack of wood and other natural resources, the societies faded and disappeared. At the other end of the spectrum, Diamond explores the sustainable practices of certain social groups and industries, providing us with role models as we confront important social choices.

Modern cities no longer have the technological tools with which to fully maintain their supply chains. New York City (NYC), for example, faces the monumental challenge of maintaining a supply of clean water for its over eight million residents. Having decided to seek an alternative to existing water-filtration technologies, the city implemented a green solution to ensure and protect its water supply. In 1997 NYC began purchasing or protecting through conservation easement what now amounts to over 280 square kilometers (70,000 acres) of forest and clean-water lakes in the Catskill Mountains of Upstate New York to ensure that they would not be subjected to industrial development and would thus remain pristine. The city also built giant aqueducts to carry water from this watershed to New York City. This supply of unfiltered water is the largest in the United States, and the water is purified as it passes through this aqueduct system. Had the city not been able to tap this source of pure water, it would have had to build a water-filtration plant, the estimated cost of which was at least $4 billion.

On its part, the State of New York sought to protect these and other lakes from degradation caused by various forms of pollution. It enacted legislation to preserve woodlands around lakes, as well as stricter laws on emissions from the fossil fuel–burning power plants that cause acid rain. State lawmakers realized that the health of huge numbers of people in NYC and elsewhere depended directly upon the protection of the state's lakes and forests.

Now that more than half of the world population lives in cities, humanity has passed a historic milestone. Cities are on the front lines in the battle to improve the health of urban environments. They are finding new allies in the upcoming generation of urban planners who understand the challenges of pollution-induced climate change as well as the benefits of the cardio-environmental model.

thirteen

WANTED: A NEW ECO-NOMIC URBANISM

It no longer suffices to change the world. We must preserve it.
GÜNTHER ANDERS, German philosopher and peace activist

TO MANY PEOPLE'S extreme frustration, mature trees are often cut down to make way for construction projects. Seeking to placate protesters and win back public support, developers are quick to promise that they will plant "new" trees. Pierre Gosselin, however, feels this approach is not the way to go; it takes at least twenty, if not forty, years before a planted tree has all of the beneficial properties of a mature tree. Gosselin, a physician and specialist in biorisk at the Institut national de santé publique du Québec (National Public Health Institute of Quebec), is in charge of planting thousands of trees throughout Quebec to reduce the adverse effects of urban heat islands, yet he believes that cities must focus on preserving existing trees and green spaces. French botanist and biologist Francis Hallé goes even further. In

Plaidoyer pour l'arbre (Plea for a Tree), he argues that a second-growth forest needs to grow undisturbed for several hundred years before it becomes an old-growth, or primary, forest and attains a state of equilibrium and sustainability.

In the meantime, the perspective that cities need their trees, green spaces, and sustainable development is slowly gaining ground. Organizations such as the United Nations, the United Nations Environment Programme, and the World Health Organization have called unanimously for the integration of green (vegetation) and blue (water drainage) corridors in urban environments as a means to protect nature, to build healthy communities, and to promote biodiversity.

Increasingly as well, city dwellers are demanding the kind of urban development in which green buildings coexist with an environment where native vegetation and biodiversity have been preserved. They have seen the negative effects of urban desertification and "mineralization"—the replacement of green organic areas by inorganic materials such as pavement—on communities and on people's health.

The Heart of the City

> *We must build cities in the countryside. The air is so much cleaner there.*
>
> ALPHONSE ALLAIS

From the point of view of healthy hearts, urban planning must emphasize and promote the following:

- Places for walking that are attractive and safe, to help residents achieve the goal of 10,000 steps per day.[1]
- Convenient and reliable public transit.
- Integration of public transit with active transit options such as walking, cycling, and skateboarding.

- Noise reduction and places of calm.
- Reductions in greenhouse gases.
- Reductions in air pollution and airborne particulates—in particular, $PM_{2.5}$ and $PM_{0.1}$, CO_2, NO_2, SO_2, VOCS, and PAHS.
- Green spaces designed to counteract the negative effects of urban heat islands.
- Restoration of green and blue corridors consistent with regional biodiversity.
- Access to fresh, healthy foods that are produced locally with a minimum of industrial additives.

From Andersen to Villadsen

For cardiologists, Denmark has once again been a source of inspiration. A Danish interventional cardiologist and researcher by the name of Henning Rud Andersen led the DANAMI 2 study that set the modern foundations of treatment for victims of heart attack. It compared the efficacy of angioplasty with traditional fibrinolysis, a medication that dissolves clots. In the DANAMI model, regional hospitals form a network with a cardiac catheterization center so that an MI patient can immediately be transferred to this center and catheterized to open the clogged artery. The study found that patients referred for dilation and stent insertion had, in the short term, better immediate outcomes and fewer complications from angioplasty and a stent than those who received fibrinolysis. In March 2010, eight years after the DANAMI 2 study results were announced, the journal *Circulation* found that in the long term the cardiac death rate was 22 percent lower in patients treated by dilation and stent insertion than in those who received the traditional intervention.[2]

In less than twenty years, interventional cardiologists had been able to offer their patients a higher quality of life and longer life expectancy.

After presenting his study at the conference of the American College of Cardiology in 2002, Henning Rud Andersen became an international celebrity, and his work has led to the restructuring and networking of cardiology centers around the world.

Another Dane has also been very influential among cardiologists, but this time the focus of the work is environmental cardiology. Kristian Villadsen, who spoke at a 2010 urban planning conference in Montreal, is an associate at the Copenhagen-based firm Gehl Architects. This firm is revolutionizing urban planning by putting people and their activities at the heart of urban development. The firm's people-first philosophy has attracted widespread international attention, and Gehl Architects now counts as clients the cities of Beijing, Dublin, and New York, among others.[3]

What is behind the new way Villadsen and others at Gehl Architects look at architecture? To begin with, they see themselves not as mere contractors but as "consultants in urban quality."[4] The philosophy of Gehl Architects is as follows:

> Gehl Architects have developed a unique working methodology based on the principle that people's priorities are the most important driver in the planning process for cities. The study of people's well-being lays the foundation for the formation of strategic planning and design work. In our project work we utilize the empirical survey and mapping methods that Professor Jan Gehl has developed, which explores the way urban areas are used.
>
> Our design solutions begin with formulating a vision and comprehensive programme of activities based on the type of life—its activities and attractions—that are inherent in a given area. The next step is to develop a public space network that can support the vision for public life both through scale, form and climate. Finally we envision how buildings can contribute to our aspirations for public life in terms of height, massing and scale as well as their ability to satisfy the need for

people-friendly functionality through proposed building uses and interaction with the public realm.

Gehl Architects' modern approach emphasizes creating sustainable environments and promoting a holistic urban lifestyle adapted to the land on which the structures are built. The design process starts with research into both the activities of the people in an area and the geography of the place itself.

The next stage is to build a community that is shaped by the pre-existing environment. This approach was used in London, England. Perhaps in the same tradition as the English garden, London was erected while preserving the area's natural watercourses, rather than filling them in or redirecting them. The natural geography had already achieved a balance between precipitation and drainage in the region, and city planners respected and maintained this arrangement. Today, the "City"—the center of London—is being redefined by new priorities: bike trails, walkways, open spaces, green belts, biodiversity. When developers try to build isolated and incongruous building projects that do not mesh with the overall character of the urban milieu, community members demonstrate their opposition.[5]

The Netherlands has several cities that may be described as avant-garde in terms of sustainable development, where high-tech companies are located in industrial complexes that are integrated into natural settings. One example is the High Tech Campus Eindhoven, originally built by the Philips group for its own research center. This complex now houses a dynamic mix of more than one hundred companies and research institutes of all sizes and specialties; among the items produced are embedded systems and microelectronic systems for health care, wellness, and personal development.

What distinguishes the High Tech Campus Eindhoven is its integration of high-tech buildings into an area of wetlands and wooded areas, and its goal of zero carbon emissions within the campus. The campus has invested over 50 million euros in its pursuit of sustainable development: natural landscaping, "green" materials, water management, recycling, energy conservation, and reduction of CO_2 emissions.

The new approach to envisioning cities is to look at everything as a whole. Rather than developing one plot of land after another in total anarchy, as we now do, we need to start with a well-structured urban plan. Is this idea radically new? Not at all. This urban planning philosophy was practiced for thousands of years in ancient cities and city states. If the process was driven by an "enlightened" tyrant, such as Pericles or Baron Haussmann,[6] the result was a masterpiece of urbanism such as the Acropolis in Athens and the great views and boulevards of Paris. When there is no oversight and development proceeds pell-mell, the end result is urban ugliness, scorned by everyone and unhealthy for people in many ways.

Are there other models of urbanism that balance quality of life, active transit, modern infrastructure, and sustainability? Yes, and, surprising as it may sound, Disneyland is a fine example.

The product of one man's vision, Disneyland embodies several principles of good urban planning. The overall design can be likened to that of an onion in which all the energy is directed toward the core. On the outside, Disneyland is connected to a vast transportation network that links it to the surrounding area and beyond. The periphery of Disneyland is a beltway of accommodations for Disneyland's visitors and staff. A comprehensive and efficient public transit system—including a monorail, buses, boats, a train, and a tramway—connects people to the heart of Disneyland, its "downtown." This is pedestrian friendly, lively, and fun, and all of the main attractions are close to public spaces, cultural areas, and concert auditoriums and theatres, as well as restaurants and spots for just relaxing. Every morning thousands of people climb aboard one or another of the site's transit services to easily travel to the heart of Disneyland, and they arrive in a good mood. Walt Disney had a vision of the type of cohesive development he wanted, and he was able to carry it out because he kept a firm hand on the reins.

The village of Mont-Tremblant in the Laurentian Mountains of Quebec is another example of modern urban planning. An internationally renowned resort, Tremblant, as locals call it, was redesigned and expanded by Intrawest on the same principles as the Disney model. Tremblant features a walkable central area that incorporates a variety of businesses and cultural activities. This core is surrounded by a band of private residences and guest accommodations with direct access to green spaces and the resort's winter and summer recreational activities, as well as to parking on the periphery. Tremblant exists in harmony with its spectacular mountainous setting largely because the developers had a vision of a community where people and nature come together.

Montreal has two similarly innovative and integrated projects, both at the site preparation stage.

Technoparc Montréal

Technoparc Montréal, a hub for pharmaceutical and technology companies, is situated close to Pierre Elliott Trudeau International Airport in western Montreal. Planned by a multidisciplinary team, the Hubert Reeves Eco-Campus project sought to create a green and sustainable work environment that would foster innovation in clean technology and energy research.[7] The aim was to have zero carbon emissions, pleasant active transit trails, and respect for biodiversity. Features of the buildings and site include ultra-modern structures designed for energy and ecological efficiency. Its wetlands and creeks are protected, wooded areas connect to regional green-blue corridors, and it has green areas for walking and open spaces. The complex brings together world-class companies and researchers in a natural setting just minutes away from the downtown area of a major North American city.

Sustainable urban design is achieved through a balance between urban planning, architecture, and landscaping. When the design process respects the natural contours and vegetation of a site as well as existing wetlands and watercourses, the outcome is an urban milieu that helps to protect air and water quality, and consequently biodiversity, while reducing the prevalence of urban heat islands. The urban-design approach of the new "Baron Hausmanns" aims to work *with* the elements of an environment rather than against them.

Outremont Campus of the University of Montreal

Though still no more than a blueprint, the University of Montreal's project for a new campus, located at a former Canadian Pacific Railway switching yard several city blocks away from the current campus, won a national urban planning award.[8] The Outremont Campus is envisioned as a university city that will serve the needs of the city of the future and be a place where the scientific principles and skills taught at the institution will find expression in the design and construction of this "test tube" university complex.

In terms of sustainability, the project will integrate a large overall area of green space, active transport, proximity to urban transit such as the metro and commuter trains, integration with the surrounding urban milieu, state-of-the-art antipollution and geothermal technologies, green roofs, zero carbon buildings, and a layout that favors optimized interactions among the researchers and professionals working there. Pedestrian corridors will link all elements of the complex.

Pitié-Salpêtrière Hospital in Paris

Such complexes are neither confined to North America nor wholly modern in concept. The Pitié-Salpêtrière Hospital, which has long prided itself on being the largest hospital in Europe (at 2,400 beds), was founded some 350 years ago. Today it sits on the same 30-hectare (74-acre) site where it first opened, but the original complex has evolved over the years and become something of a medical Disneyland. The early historic buildings are now interspersed with new pavilions and research centers, as well as the buildings of the Faculty of Medicine. But because the site is so large, all of the buildings, including the ultramodern Institute of Cardiology and Babinski Building for neurology, are surrounded by beautiful and spacious grounds. Public transit—two railway lines and three subway lines—link the complex to Paris and beyond. Here is an example of a facility that has evolved to meet the needs of both the environment and medicine.

Some urban renewal projects truly do bring renewal. The restoration of the Pins-Parc traffic interchange in Montreal was such a project. Cement overpasses were replaced by city avenues bordered by vegetation and pedestrian and bike paths. Vehicle traffic now flows smoothly, a welcome change from the frequent traffic jams of before. The restoration, which reduced the surface area of concrete and asphalt, also mitigated an urban heat island.

Health Tree Day

The medical, nursing, and technical support staff at the Laval health complex Cité de la santé have celebrated Health Tree Day in September since 2007. In honor of the theme of health and the environment, and in conjunction with National Tree Day, since its founding in 2011, they have voluntarily planted hundreds of trees around the center's health and social services (CSSS) buildings. Those behind the annual initiative have adopted the motto "We are rooted in our community" and opened up the program to local community organizations, with the result that early-childhood centers and long-term and palliative care centers in Laval now also celebrate Health Tree Day. By planting trees and engaging in other activities aimed at reducing the environmental footprint of these organizations, the volunteers help improve the quality of life and the environment.

In 2012 the remaining CSSSS in Montreal, Sherbrooke, Trois-Rivières, and Laval joined the movement. The objective is to eventually include all of the CSSSS in Quebec. Incidentally, cities aren't the only beneficiaries of sustainable planning practices. Many people are working to reverse the damage to rural land from deforestation. One such volunteer is Marcel Leboeuf, a popular Quebec actor who spends a lot of time and energy restoring forested habitats. He created an organization called the Eastern Townships Forest Research Trust[9] with Benoit Truax, who holds a PhD in environmental science, and Daniel Gagnon, professor of plant ecology at the Université du Québec à Montréal. They consult with agricultural producers and other private land owners on how to restore areas, such as the banks of streams and rivers, that have sustained heavy ecological damage from deforestation. They have also planted many thousands of trees, particularly butternut, white pine, and red oak, valuable species that once flourished in the Eastern Townships region of Quebec and the northeastern United States but are now rarely seen. These protectors and planters of trees know that forests provide us with many different benefits.

fourteen

THE RED AND THE GREEN

CORONARY TREE. *Aortic root. Common arterial trunk. Right bundle branch.*

Anyone who studies the heart cannot help but be amazed at the extent to which cardiology has borrowed concepts and terminology from botany. Medical scientists inherited a tradition of nomenclature begun by our distant ancestors, who described plants in order to harness their properties. Naming is the first step in understanding. When we name a child, for example, we confer an identity on her or him. She is no longer "the baby"; she is "Audrey." Perception becomes reality.

Vascular arborization. Mitral leaflet. Marginal branch. Capillaries.

The botanical terms found in cardiac anatomy are more than just metaphorical descriptions of the cardiovascular system. Similarly, the evolutionary connection between the circulatory systems in humans and trees goes beyond symbolism. Nature devised parallel and complementary formulas to create the kingdom of plants and the kingdom of animals. These two kingdoms have, in fact, the same ancestor, and the process that resulted in their differentiation began more than 600 million years ago. Because they have many structures in common, these two kingdoms have continued to co-evolve and shape each other ever since. The discovery of these parallels is an ongoing source of marvel.

The vital connection between animals and plants has endured through the eons because of two proteins: hemoglobin and chlorophyll—the red and the green. They each have the same mission: to ensure the exchange of oxygen and carbon dioxide between plants and animals in the world's most ancient barter.

Green chlorophyll is the essential component in that most amazing photovoltaic battery: the world's plant cover, which performs the miracle of transforming carbon dioxide into oxygen, a fundamental element of life—and of hearts being treated for respiratory distress. Trees, the primary source of atmospheric oxygen, are thus the crucial allies of human beings, in general, and of cardiologists in particular.

Oxygen has not always been plentiful in Earth's atmosphere. At the beginning of geologic time, perhaps 4.5 billion years ago, most of the carbon and oxygen atoms on the new planet were bound together as carbon dioxide. Free oxygen became abundant only after blue-green bacteria and algae appeared and began the crucial process of photosynthesis, about 3.5 billion years ago. After another billion years, oxygen was plentiful. Fast-forward to 400 million years ago, when the first trees appeared. The next 100 million years was the age of trees, during which most of the Earth's fossil fuels were created from the wealth of decaying plant life. During that period, oxygen became even more plentiful in air than it is today—a perfect environment for animals to evolve and flourish.

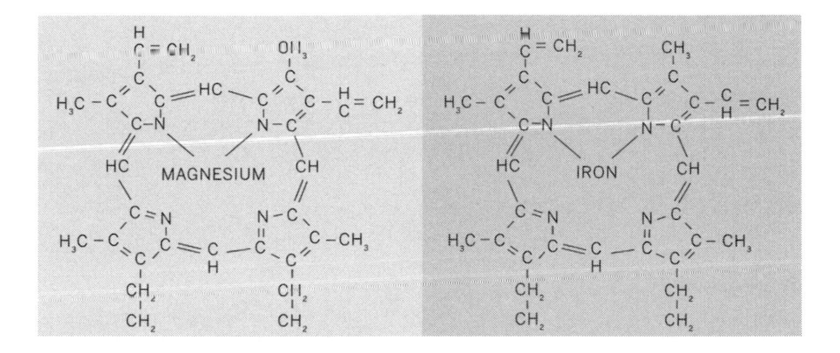

FIG 4 *Chlorophyll and hemoglobin.*

Today, only about 9.4 percent of the Earth's total surface is covered by trees, but this area nonetheless represents enough trees to produce, every second, the tons of oxygen needed to ensure that the atmosphere sustains life on Earth—namely, that it maintains its oxygen at around 21 percent. The energy needed for this vital exchange (O_2-CO_2) comes from the sun, which has a life expectancy of five billion years—and so is fully sustainable as far as humans are concerned.

Hour in, hour out, the sun supplies Earth with a thermal energy equal to that of burning 21 billion tons of coal. And the sun is ultimately the source of the energy that powers every one of the 100 trillion cells in our bodies. In a process complementary to photosynthesis, our cells metabolize and burn the nutritional materials first created by plants. Blood transports hemoglobin, the carrier of oxygen, to all parts of the human body, while the sap of the tree both brings water to the chlorophyll that resides in the leaf cells and carries away simple carbohydrates to the roots and growing parts of the tree.

At the core of each of these proteins is an atom of a cherished metal—iron in one and magnesium in the other. Iron gives hemoglobin its red color. Magnesium is responsible for the green in chlorophyll. But, and this is really interesting, the protein structures that surround these two metal atoms are *identical*, even though each is used by a very different kingdom—iron by animals and magnesium by plants (Figure 4). They are the same molecule! This surprising and very close relationship between human beings and trees is molecular in origin.

In the capillaries of the lungs, red blood cells come into contact with gases in the alveoli of the lungs. The hemoglobin opens its protein branches and its iron atom picks up an oxygen atom. Thus oxidized, the hemoglobin brightens, changing from the deep purple color of venous blood to the beautiful rich red of arterial blood.

The molecule of hemoglobin transports its atom of oxygen to the body's cells that hungrily wait for it. The cells effortlessly absorb this oxygen and use it to produce the substance that fuels our cells, adenosine triphosphate (ATP). The same "soft" combustion reaction also produces carbon dioxide, which is returned first to the lungs and then to the atmosphere.

As we have seen, carbon dioxide is absorbed by the leaves of trees and other plants. The magnesium atom in each packet of green chlorophyll allows sunlight energy to be captured and transported to the intracellular sites of photosynthesis, where carbon dioxide and water are combined into sugar and oxygen. The fundamental circle of life is thereby closed.

An axiom of the way humans perceive color is that red excites and green calms. The red in blood is a powerful pigment; just one drop of blood tints an entire liter of water. Blood appears instantly on the skin after the smallest scratch, a highly efficient alarm signal. With blood comes pain, or so we believe, and at the sight of blood we take action to protect ourselves and to stop the bleeding. In some people the mere sight of blood sets off vasovagal shock, which slows the pulse and reduces blood flow, thus minimizing the loss of blood at the wound.

Nature chose red, the most alarming of all the colors, to alert us to the loss of even the smallest amount of our precious blood.

Coincidence or necessity? In contrast, green is considered a relaxing color. To painters or anyone who uses imagery to create poetry, red and green are complementary and mutually enhancing colors. This concept is supported by optical scientists as well as by biologists, who first discovered the complementary actions of hemoglobin and chlorophyll.

We are slowly gaining a better understanding of the symbiotic relationship between the health of plants and the health of humans. However, in this mutual dependency, humans alone have the capacity to respond to save their allies. We have to act to protect and sustain our environment—above all its trees.

This is the profound message of *The Man Who Planted Trees*.

fifteen

THE HEART OF THE TREE

We are at war with Nature.
If we win, we are lost.
HUBERT REEVES

MONEY DOES NOT grow on trees. They have so much more to offer.

Our time, which many call the anthropocentric era or the Neocarboniferous era, is emerging as a post-industrial age. The threat of climate change; the degradation of air; the increase in smog, urban heat islands, desertification, and extreme storms and weather events; as well as the degradation and depletion of fresh water and forested areas are altering the central focus of *Homo sapiens*, who are turning their technology toward environmental preservation.

One sign of this change in direction is that activities that harm Earth's ecosystem are now often illegal. We are finally acknowledging the extent of ecological harm to the planet and are collectively becoming more receptive to the needs of the environment and to the idea that nature—Earth's biosphere—has rights. This attitude is reflected in the emergence of the new global legal specialty: the law of environmental responsibility.

One public body involved in the field is SERDEAUT: Sorbonne Études et Recherche en Droit de l'Environnement, de l'Aménagement, de l'Urbanisme et du Tourisme (Sorbonne studies in research on the law of environment, development, city planning, and tourism) at Paris 1 Panthéon-La Sorbonne University. An example of the work being done was the November 2008 report by assistant professor and lawyer Marta Torre-Schaub on the legal impacts of ecological damage, in which she examined the possibility of envisioning a global concept of ecology that would make polluting wrong or even criminal.

The twenty-first century has already pulled away from the twentieth century as a result of the growing consciousness that, first, achieving quality of life means we have to conserve the environment rather than simply subjugate it and, second, the relationship of humans and nature is one of equality and reciprocity. This transformative notion is thoroughly explored by David Suzuki in his milestone work *The Sacred Balance*. Suzuki proposes ways for us to reconcile our quality of life and our survival—a balance threatened by the manic race for "progress" that we have been engaged in for the past hundred years. We have now caused so much damage, out of either ignorance or arrogance, that we have been forced to globally re-evaluate the balance between our needs and our natural resources.

The central events of the first decade of the twenty-first century were the economic crisis and the environmental crisis. Together they constituted "a perfect storm," according to the economist L. Jacques Ménard, who made this point in a February 2009 letter to the editor in Montreal's *La Presse* newspaper. Whether or not this convergence was mere coincidence, some observers have interpreted it as an opportunity to bring together the efforts to address these two crises and create new opportunities for individ uals, for society, and for industry.[1]

For individual people, ecology is like gardening: repeating thousands of small daily gestures in an organized framework, not unlike one's daily hygiene regimen. We learn by doing, and daily gestures become as satisfying as they are productive. For society, ecology is all about bringing simple opportunities within the reach of individuals in their daily spheres. For industry, it is a matter of stepping up to a new challenge: to provide for consumers products with an environmental footprint no larger than that of nature's.

Not so long ago the ecological degradation or economic distress of a distant Asian or European nation affected North Americans very little or not at all. Now, thanks to globalization, it does affect us, and very directly. By adapting to the principles and long cycles of Mother Nature, we may find a new approach that reconciles both our economic and ecological needs.

Various universities and engineering firms are in the final stages of developing technologies to clean up the environment. We are creating systems to purify and store water; to cool and heat homes using more energy-efficient technologies; to reduce urban smog; to clean our atmosphere of oxidative particles and carbon dioxide; and to sequester toxic airborne, aquatic, and terrestrial pollutants to mitigate the effects of ultraviolet rays and the risk of skin cancer. We are conceiving new air conditioning and heating methods that draw on immediately available energy sources—wind, sun, tides, and rivers—thus avoiding the need to physically transform energy sources, by refining, for example. Technological enhancements in geothermal energy, photon capture, wind energy, tidal power, and geothermal heat pumps are on the way, along with water filtering and purifying systems that use sedimentation chambers and walls that block noise and wind. Some simple measures are already at hand, such as the use of awnings, canopies, and sunscreen to protect skin from the damage caused by ultraviolet rays (the product of our wanton depletion of the ozone layer). In addition, these new technologies integrate indirect benefits that may promote well-being, lower stress, and stimulate healthy lifestyle choices, thus creating a favorable environment for maintaining health instead of degrading it. Safeguarding the environment, improving the workplace, and enhancing quality of life are major drivers in current urban planning and architecture. But if you think about it, despite all of these efforts, all this research and reliance on a technological Holy Grail, the solution is right there in front of us.

It is called a tree.

It seems to me that what we need to do to ensure human health and survival is to reappropriate the tree. Trees are, in many respects, our yang, our direct complement, our main interface with nature. And let's not forget that trees and humans must together inevitably face the grim reaper known as the sixth extinction. The International Union for Conservation of Nature has estimated that, if the deterioration of the environment and biosphere continues at the present rate, mammals weighing over 3 kilograms (6.5 pounds) and all large trees will become extinct.[2] Trees and people, both will expire in a battle launched by people who thought they had the right answers. We human beings need to shed our sense of self-importance and learn to protect our partners of long standing, the trees, or both species will perish. For as we must certainly realize, trees can manage just fine without us; the contrary is not, however, true.

The first line of evidence: humans suck in oxygen and exhale carbon dioxide; trees capture carbon dioxide and emit oxygen. This relationship is symbiosis at its finest. When, for example, it comes to treating respiratory distress, a condition in which patients experience oxygen desaturation and retain carbon dioxide, cardiologists can find no better allies than trees because they furnish oxygen and suck up carbon dioxide. But trees offer other and sometimes surprising benefits in addition to those associated with their capacity for the vital exchange of oxygen and carbon dioxide (O_2-CO_2) molecules.

The Man Who Planted Trees

The 1987 Academy Award-winning short animation film *The Man Who Planted Trees* (French title, *L'homme que plantait des arbres*), by Frédéric Back of Montreal, immortalized the eponymous poetic tale by French writer Jean Giono, first published in 1953. In images evocative of the paintings of Wassily Kandinsky, the film depicts the heartwarming story of Elzéard Bouffier, whose chosen purpose in life is simple: to plant trees.

Upon first encountering this story, we can't help but be moved by its beauty, tenderness, poetry, and humanity. However, in the light of what we now know about global ecosystems, re-reading it adds reason to our emotion and amplifies both. Like the foundational books of the world's great religions, Giono's text is layered with different interpretations and meanings, from the simple to the transcendent, as was his intention in writing it. Perhaps the greatest attribute of the story, however, is its universality.

Jean Giono shows that the simple act of reforesting one region of the planet carries with it implications that evoke deep philosophical questions, as you will discover in reading several excerpts from this great work over the next few pages.[3]

> For three years he had been planting trees in this wilderness. He had planted one hundred thousand. Of the hundred thousand, twenty thousand had sprouted. Of the twenty thousand he still expected to lose about half, to rodents or to the unpredictable designs of Providence. There remained ten thousand oak trees to grow where nothing had grown before.

In the eighteenth century, before the world's forests were pillaged during the next two centuries, François-René de

Chateaubriand wrote, "*Forests* precede civilizations and deserts follow them," and, indeed, the phenomenon of deforestation is not new. People have always cut trees to make space for towns and cities, to free up agricultural land, and to use the wood for building and heating. In the past two hundred years, two-thirds of the world's forests have been severely reduced. An estimated 50 percent of forests have been eradicated worldwide. The landmass of Europe has gone from being 90 percent wooded to less than 15 percent wooded. The cutting of forests has led to the destruction of habitats and the extinction of thousands of living species worldwide. One billion people and one-quarter of Earth's landmass are threatened. Across the planet, rivers and lakes are drying up.

Apart from its role in species loss, deforestation is responsible for 20 percent of atmospheric carbon dioxide emissions, which exceeds the total CO_2 emitted from combustion by all forms of transportation. A study titled "The Economics of Ecosystems and Biodiversity" presented in 2008 at the United Nations Convention on Biological Diversity has estimated that forest destruction will cost 2 trillion euros annually, or 6 percent of global gross domestic product.[4]

Deforestation leads to soil depletion and erosion. From Beijing to North Africa, severe sand storms pummel cities and pave the way for desertification. The eradication of forests leads to excess stormwater runoff as well as catastrophic flooding and deadly mudslides.

> His name was Elzéard Bouffier. He had once had a farm in the lowlands. There he had had his life. He had lost his only son, then his wife. He had withdrawn into this solitude where his pleasure was to live leisurely with his lambs and his dog. It was his opinion that this land was dying for want of trees.

In 1992 a new movement was born, ecopsychology, which focuses on the relationship between humans and the environment.[5,6] Recall that psychology initially focused on the individual and then, with the advent of psychoanalysis, on the collective unconscious, and more recently, with the advent of systems therapies, on emotional and social relationships. Today psychology includes the notion that humans derive significant psychological benefits from the natural environment.[7]

In hospitals the ability to look out on nature reduces the incidence of postoperative complications as well as the convalescence period for patients. Patients recuperating from surgery in a room with a view of nature spend less time in hospital, have fewer negative evaluations from staff, take fewer analgesics for pain, and have somewhat lower scores for minor postoperative complications than do patients who see nothing out the window but a brick wall.[8]

In cities, hedges and other types of greenery surrounding homes help improve social relations among neighbors, and life is calmer in neighborhoods with access to green spaces.[9]

Other studies have shown that proximity to a natural environment is correlated with a decline in crime, aggression, and violence, and, in contrast, with a rise in civic awareness and improved relations among neighbors.[10,11] Incorporating the natural world in communities through activities that encourage outdoor participation can increase feelings of belonging to a community and reinforce urban social ties.[12,13]

L.M. Wilson demonstrated in 1972 that patients whose hospital rooms lacked a window suffered from significantly more delirium than those who had a window.[14] Contact with nature can improve concentration and reduce symptoms of attention-deficit disorder in children. It can also help improve students' self-discipline and emotional development. Contact with or views of an outdoor natural area can reduce indicators of stress such as skin conductance and high blood pressure and increase the sense of well-being and mental health in patients.[15]

Whole civilizations have disappeared as a result of deforestation, including, as Jared Diamond has described so compellingly in *Collapse*, the one on Easter Island. There are now signs that deforestation may be having effects on a planetary scale. Globally, the level of atmospheric carbon dioxide has already risen beyond 350 ppm (parts per million), the limit that many regard as safe. It is currently about 400 ppm. This alarming fact explains why climate-change experts consider forest renewal to be an essential component of the strategy to reduce atmospheric carbon dioxide to safe levels. But our expectation that global ecosystems will save us by absorbing the carbon currently in the atmosphere will prove misplaced unless we drastically reduce deforestation and also the emissions generated by fossil-fuel power plants, vehicles and other modes of transportation, as well as agriculture and animal husbandry.

So important is the role of forests to the alleviation of climate change that the United Nations Environment Programme launched an initiative called REDD (Reduction of Emissions from Deforestation and Forest Degradation) to create mechanisms to ensure, and financial incentives for, the preservation of existing forests. An international partnership for the protection of tropical forests was officially launched in May 2010 in Oslo; it brought together nine donor nations (Norway, the United States, France, Germany, the United Kingdom, Australia, Japan, Sweden, and Denmark), the European Union, and about forty forested countries.[16]

> I told him that in thirty years his ten thousand oaks would be magnificent. He answered quite simply that if God granted him life, in thirty years he would have planted so many more that these ten thousand would be like a drop of water in the ocean.

The United Nations Environment Programme also set up the Plant for the Planet Foundation's Billion Tree Campaign. It was inspired by a Kenyan citizen action campaign called the Green Belt Movement led with great courage and determination by Wangari

Maathai, who in 2004 was awarded the Nobel Peace Prize for her work. Since its beginnings in 1977, this movement has planted more than 30 million trees in Africa.

At the time of its launch in 2006 the mission of the Plant for the Planet campaign was, obviously, to plant one billion trees. But the project has far exceeded expectations: in 2007 alone more than 1.7 billion trees were planted, and during the first five years of the campaign over 12.5 billion trees were planted.[17]

> It was with no other objective that I again took the road to the barren lands... Since the day before, I had begun to think again of the shepherd tree-planter. The oaks of 1910 were then ten years older and taller than either of us. It was an impressive spectacle... He had pursued his plan, and beech trees as high as my shoulder, spreading out as far as the eye could reach, confirmed it.

Reforestation promotes carbon sequestration and water retention. A forest is a retention basin, a vertical lake, in which a mature tree can hold up to 400 liters of water. Root development slows runoff and improves soil drainage and the flow of groundwater to the aquifer. The forest canopy condenses the humidity in the air and regulates rainfall, both to the benefit of streams and lakes.[18] Mature trees are crucial vectors for biodiversity. Each tree is a small planet that hosts an immense variety of plants and animals. The roots develop a symbiotic relationship with underground fungi that decompose the soil's nutrients, preparing them for absorption by the tree's fine roots. Colonies of insects set up shop amid the roots, aerating and composting the soil and pollinating nearby plants. Birds shelter in the foliage. Small mammals find food and shelter on the tree, sometimes right inside it. Around its perimeter, other plants root and grow.[19]

Despite the large number of trees planted by Plant for the Planet, whose motto is "Stop talking. Start planting," and other campaigns such as Global ReLeaf (at time of writing, more than 44 million trees planted), we cannot sit back, content to rest on our laurels. Each year 13 million hectares (32 million acres) of forest are cut. To replace all of the trees that have disappeared in the last decade alone, we would have to reforest 130 million hectares (320 million acres).[20]

> Everything was changed. Even the air. Instead of the harsh dry winds that used to attack me, a gentle breeze was blowing, laden with scents. A sound like water came from the mountains: it was the wind in the forest. Most amazing of all, I heard the actual sound of water falling into a pool...
>
> The new houses, freshly plastered, were surrounded by gardens where vegetables and flowers grew in orderly confusion, cabbages and roses, leeks and snapdragons, celery and anemones. It was now a village where one would like to live.

Children who live in communities with a large number of parks, green spaces, and outdoor recreational facilities are more likely to engage in walking and cycling (active transit), according to a study presented at the 2009 conference of the American Heart Association by Tracie Ann Barnett, assistant professor in the Department of Social and Preventive Medicine at the University of Montreal and researcher at Sainte-Justine University Hospital Research Centre. Working in collaboration with the Quebec Adipose and Lifestyle Investigation in Youth (QUALITY) study, which collected data on 600 children and their biological parents in their homes to determine the natural history of the children's cardiovascular disease risk profile, Tracie Barnett found that for each additional park within a radius of 750 meters (820 yards) from a child's home, the probability that girls would walk to school was twice as high, and the probability that boys would go for a walk to relax was 60 percent higher.[21] She concluded that there is a strong correlation between walking and physical activity of all types, and the number of nearby public open spaces and recreational areas, in particular, parks, playgrounds, and sports facilities.[22] Urban forests promote the health of the communities in which they are located and also that of the people who live near them. They play a significant role in reducing energy needs (heating and cooling costs) and in the absorption of atmospheric carbon dioxide. According to Tree Canada, a single urban tree can save five to ten times the overall carbon of a tree in a rural area. One mature tree can produce enough oxygen to meet the needs of four people for an entire day. A tree can reduce airborne dust particles by as much as 7,000 particles per liter (or quart) of air, and absorb 22 kilograms (49 pounds) of carbon dioxide per year. The vast expanse of the tree's foliage filters aerosol particulates and the pollutants from fossil fuels.[23]

One hectare (2.5 acres) of urban forest, or its vegetative equivalent planted in an urban milieu, can eliminate 15 tons of smog per year, or the equivalent of the emissions of 75 cars over the same period. When the outdoor temperature rises to 18 degrees Celsius (64 degrees Fahrenheit) or higher, air pollution is aggravated. These atmospheric pollutants combine with sunlight to produce even more hazardous secondary compounds, such as ground-level ozone. An additional benefit of wooded areas is that, when air passes through them, many of the volatile secondary compounds are absorbed by leaves and physically decomposed into less hazardous substances. Trees are powerful air cleaners: they absorb and metabolize oxidized organic trace gases.[24]

By providing wind buffering and shade, trees can reduce the demand for and thus the cost of energy for heating and air conditioning. Homes protected from wind by trees can save from 10 to 15 percent of the cost of heat and up to 30 percent of the cost of air conditioning.

Numerous studies and analyses of literature from Australia, Japan, the Netherlands, Norway, Sweden, the United Kingdom, the United States, and Canada, among others, have shown how trees, woodlands, and green spaces promote people's health and well-being. These studies demonstrate the benefits of access to not only activities in nature, such as walking and biking, but also to views of nature—for example, through windows or in the course of one's daily activities. One consequence of this research was that in 2005, in Great Britain, the government department responsible for forestry signed a health agreement with the country's public health agencies to promote the use of the outdoors for public health.[25]

Can the health benefits of a green community be measured in relation to cardiovascular health?

Heart of a Tree and Heart of a Human

The 2008 publication of the article "Effect of Exposure to Natural Environment on Health Inequalities" in the prestigious medical journal *The Lancet* seemed to confirm the growing interest in, and validity of, the cardio-environmental model. This study found that cardiovascular mortality was lower in populations with higher incomes and greater exposure to green space compared with more deprived populations with less exposure to green space. In other words, cement-encased city centers can be lethal, especially for low-income residents.[26]

The article was written by Richard Mitchell, a professor in the Public Health and Health Policy Department at the University of Glasgow, and Frank Popham, a professor at the School of Geography and Geosciences at the University of St. Andrews. Once again an epidemiologist and a geographer had joined forces to carry out groundbreaking research—in this case a five-year survey of mortality data based on a sample of a whopping 40 million people in England (made possible because of computers and databases, of course).

The authors' hypothesis: if green spaces are positively correlated with better health, this link should be reflected in a difference in mortality between those living in green environments and those living in concrete neighborhoods.

They also tested the hypothesis that income disparities are associated with different life expectancies—in other words, that the rich live longer than the poor. Although this conjecture has been accepted as true for as long as differences in caste or class have existed, Mitchell and Popham's research shed new light on the concept in that it examined the interaction of the two variables of income and green space—that is, the environment in which people live.

They wrote: "Our hypothesis was that income-related inequality in health would be less marked among populations with greater exposure to green space because it has the potential to modify pathways via which lower socio-economic position can lead to disease." Thus they expected the green benefits to reduce some of the traditional differences in health between rich and poor.

Using data on the entire active adult population of England, excluding retired and elderly persons, the authors categorized the Queen's 40 million subjects along two axes: (1) rich and poor, into four groups, or quartiles, and (2) living in the most concreted-over environment compared with the greenest environment, into five groups, or quintiles. From mortality data they found that 366,000 subjects had died between 2001 and 2005. These subjects were then categorized according to income and degree of exposure to a green milieu.

The authors' two main conclusions were as follows:

1. Overall mortality, low- and high-income persons taken together, was 6 percent lower for those who lived in the greenest milieu. If this result seems very familiar, it is; the Harvard Six Cities Study had similar findings.
2. Among those living in the least green environment, the relative risk of a cardiovascular death between low- and high-income persons was 2.19. In other words, low-income inner-city dwellers had an added risk for cardiovascular death of 119 percent over the high-income residents.

Among those living in the greenest milieu, the relative risk of a cardiovascular death was 1.54 for the low-income residents, meaning that the poor had an added risk of 54 percent over their richer neighbors—a much lower added risk than if they had been living in a non-green neighborhood.

Thus, in the greenest environment, the extra risk of cardiovascular death due to being poor is *less than half* of the extra risk in the concrete jungle. Or, in public health terms, no known medical intervention can have as great an impact as a green environment in reducing the discrepancy in cardiovascular health between rich and poor. The implications of these findings are clear: a healthy living environment for everyone is necessary if we are to reduce the social inequalities of disease.

When I met Mitchell in his magnificent, verdant office at the University of Glasgow, he was more philosophical than categorical when it came to explaining this association between green space and health, and especially regarding the huge gap between rich and poor in cities.

In relation to the hypotheses of the study, he saw a link between "green space" and "lack of pollution" and the opposite, "lack of green space" and "pollutants"—particularly from gasoline and heating oil. In the case of green space, he perceived the influence of positive psychological engagement, an almost universal human response to nature, and one associated with lower stress. He also thought it likely that the trees' active and passive filtering of the air protects people from airborne pollutants.

The Emerald Ash Borer: The Death Carrier

The emerald ash borer, a deadly parasite that is ravaging ash trees in the eastern and midwestern United States and Canada, is an unfortunate consequence of globalization. This insect has, however, provided epidemiologists and botanists with another opportunity to determine whether trees are a protective factor for heart health. Originally from Asia, the emerald ash borer was discovered in 2002 in Detroit, Michigan, and is an invasive species introduced to North America either inadvertently or through negligence. It has been aggressively destroying ash trees in the Great Lakes area ever since and has now been found in Toronto and Montreal; ash trees, for example, make up 20 percent of Montreal's urban canopy. There is no really effective way to eradicate the beetle, which has already killed an estimated 100 million infested trees in North America.

This tree parasite became the unlikely star of a huge epidemiological study in which scientists set out to assess the overall health impact and in particular the cardiovascular impact of this massive kill-off of trees. The results of the study are startling. In regions of the fifteen U.S. states where the ash borer has been most active, the authors found there were 6,000 more deaths from pulmonary disease and 15,000 more deaths from cardiovascular disease as compared with uninfested areas. Because the insect itself is not directly toxic to humans, the authors concluded that these deaths must be the result of the loss of trees.[27]

How exactly do trees positively affect cardiovascular health? For an explanation let us follow the path of *shinrin-yoku* and immerse ourselves in a branch of Japanese Zen philosophy.

Shinrin-Yoku

Humans appreciate the environment of the forest for its calm atmosphere, temperate climate, and fresh, scented air, all of which promote relaxation and help to alleviate stress. Shinrin-yoku, whose literal meaning is "bathing in the forest," is the name of a form of relaxation widely practiced in Japan. As the website of California's Sonoma County Shinrin-yoku Coalition tells us, "Go to a Forest. Walk slowly. Breathe. This is the healing way of Shinrin-yoku, the medicine of simply being in the forest."[28] Shinrin-yoku, it is interesting to note, is not an ancient Japanese practice but a relatively modern one; the term *shinrin-yoku* was coined in 1982 by the Japanese Ministry of Agriculture, Forestry, and Fisheries.

The Message Tree

The benefits of shinrin-yoku come not just from the stillness and beauty of the forest but also from the volatile essential oils emitted by trees, the subtleties of which we are only now beginning to understand because, although it is likely that many volatile molecules have positive biological effects, the precise nature of these impacts has yet to be determined.

People tend to think that medications delivered in liquids, lozenges and tablets, or injections are more effective than volatile substances. But recall that 20 kilograms (44 pounds) of air pass through the lungs per day, and note that breathing pharmacological and other plant-based inhaled substances (tobacco, cannabis, cocaine, morphine, and so on) and anesthetizing gases can indeed have a potent effect.

Teams of physiologists are working to understand the mechanism by which a walk in a forest produces a feeling of well-being in people. It may have something to do with a concept revealed by biochemists—that, for millions of years, trees have exchanged messages using airborne and root-transmitted proteins.

In much the same way that scientists discovered the vitamins in fruits and vegetables, they have identified a number of volatile organic compounds (VOCs) from trees that have a physiological effect on humans. These airborne compounds contain phytoncides, or essential wood oils, such as alpha- and beta-pinene, 1.8-cineole, D-limonene, and humulene, which may have antibacterial, anti-inflammatory, and even anticarcinogenic properties. Studies are underway to determine how these substances work.

Researchers now have evidence that exposure to these airborne molecules from trees can reduce arterial contraction; lower arterial blood pressure; reduce the release of stress hormones such as cortisol, adrenaline, and norepinephrine; lower such factors of vascular inflammation as C-reactive protein; and increase the number of white blood cells that defend against viruses ("natural killer" cells).[29]

A large Japanese study observed participants from twenty-four Japanese communities and compared their clinical and biological parameters before and after they walked in both a forest and in a city area. The study found that forest environments are linked with lower inflammatory factors and concentrations of cortisol and, especially, with lower blood pressure than are city environments and that the extent of these effects are comparable to those of an antihypertensive agent.[30]

So there you have it: taking a regular walk in the woods can help you avoid the need to take medication.

By way of an epilogue, here is an excerpt from Muriel Barbery's *The Elegance of the Hedgehog:*

> After I'd had a chance to think about it for a while I began to understand why I felt this sudden joy when Kakuro was talking about the birch trees. I get the same feeling when anyone talks about trees, any trees: the linden tree in the farmyard, the oak behind the old barn, the stately elms that have all disappeared, the pine trees along windswept coasts, etc. There's so much humanity in a love of trees, so much nostalgia for our first sense of wonder, so much power in just feeling our own insignificance when we are surrounded by nature... Yes, that's it: just thinking about trees, and their indifferent majesty and our love for them teaches us how ridiculous we are... and at the same time how deserving of life we can be, when we can honor this beauty that owes us nothing.[31]

Notes

CHAPTER ONE

1. "U.S. Markets for Interventional Cardiology Products." Life Science Intelligence 2012. http://www.lifescienceintelligence.com/market-reports-page.php?id=A208. Accessed May 2, 2013.
2. "Cardiovascular Diseases (CVDs): Fact Sheet No. 317." World Heath Organization. http://www.who.int/mediacentre/factsheets/fs317/en/index.html. Updated March 2013.
3. John D. Cantwell. "Cardiovascular Disease and Olympic Games in China." *American Journal of Cardiology* 101 (2008): 542–43.
4. "History of the Framingham Heart Study." Framingham Heart Study. http://www.framinghamheartstudy.org/about/history.html. Updated April 24, 2013.
5. R.D. Fraser. "Average Annual Number of Deaths and Death Rates for Leading Causes of Death, Canada, for Five-Year Periods, 1921 to 1974. Series B35–50." *Section B: Vital Statistics and Health.* Statistics Canada. http://www.statcan.gc.ca/pub/11-516-x/sectionb/4147437-eng.htm. Updated October 22, 2008.
6. "About the Framingham Heart Study Participants." Framingham Heart Study. http://www.framinghamheartstudy.org/participants/index.html. Updated April 24, 2013.
7. "William B. Kannel, MD, Pioneer in Cardiovascular Epidemiology, 1923–2011." Framingham Heart Study. http://www.framingham heartstudy.org/participants/othernews.html. Updated April 24, 2013.

8. The Canada Gairdner International Award is given annually at a special dinner to three to six people for outstanding discoveries or contributions to medical science. Receipt of the Gairdner is traditionally considered a precursor to winning the Nobel Prize in Physiology or Medicine; as of 2007, sixty-nine Nobel Prizes have been awarded to prior Gairdner recipients. Many people believe that Kannel also deserved to win the Nobel Prize for his great gift to humanity, but he died before this honor could occur. See http://www.gairdner.org/content/william-b-kannel and http://en.wikipedia.org/wiki/Gairdner_Foundation.
9. Visit http://genome.ucsc.edu/cgi-bin/hgTracks?db=hg19.
10. "Welcome to the Okinawa Centenarian Study!" Okinawa Centenarian Study. http://www.okicent.org. Accessed May 2, 2013.
11. "Okinawa, les secrets du record mondial de longévité." AgoraVox. http://mobile.agoravox.fr/actualites/sante/article/okinawa-les-secrets-du-record-18187. Posted January 26, 2007.
12. World Health Organization. *Atlas of Health in Europe*, 2nd edition. Geneva: World Health Organization, 2008.
13. World Health Organization. *Atlas of Heart Disease and Stroke*. Geneva: World Health Organization, 2004.

CHAPTER TWO

1. T.R. Dawber, G.F. Meadors, and F.E.J. Moore. "Epidemiological Approaches to Heart Disease: The Framingham Study." *American Journal of Public Health* 41 (1951): 279–86.
2. A. Allan et al. "Computed Tomographic Assessment of Atherosclerosis in Ancient Egyptian Mummies." *Journal of the American Medical Association* 302 (2009): 2091–93.
3. M. Gurven, et al. "Inflammation and Infection Do Not Promote Arterial Aging and Cardiovascular Disease Risk Factors among Lean Horticulturalists." *PLOS ONE* 4, no. 8 (2009): e6590. Available at doi:10.1371/journal.pone.0006590.
4. Ibid.

CHAPTER THREE

1. Suzuki urges us to restore the "sacred balance" between ourselves and the biosphere. David Suzuki, with Amanda McConnell and Adrienne Mason. *The Sacred Balance*, 3rd edition. Vancouver: Greystone Books, 2007.

2. A.J. Werner. "The Death of Mozart." *Journal of the Royal Society of Medicine* 89 (1996): 59.
3. David Levy. "Gustav Mahler and Emanuel Libman: Bacterial Endocarditis in 1911." *British Medical Journal* (Clinical Research Edition) 293, no. 6562 (1986): 1628–31.
4. Andy Wachowski and Lana Wachowski. *The Matrix*. Directed by Andy Wachowski and Lana Wachowski (Warner Bros., 1999).

CHAPTER FOUR

1. Rémi Cadet. *L'invention de la physiologie – 100 expériences historiques.* Paris: Bélin, 2008.

CHAPTER FIVE

1. Lenny R. Vartanian, Marlene B. Schwartz, and Kelly D. Brownell. "Effects of Soft Drink Consumption on Nutrition and Health: A Systematic Review and Meta-Analysis." *American Journal of Public Health* 97 (2007): 667–75.
2. Julie R. Palmer et al. "Sugar-sweetened Beverages and Incidence of Type 2 Diabetes Mellitus in African American Women." *Archives of Internal Medicine* 168, no. 14 (2008): 1487–92.
3. Matthias B. Schulze et al. "Sugar-sweetened Beverages, Weight Gain, and Incidence of Type 2 Diabetes in Young and Middle-aged Women." *Journal of the American Medical Association* 292 (2004): 927–34.
4. L. Chen et al. "Prospective Study of Pre-Gravid Sugar-sweetened Beverage Consumption and the Risk of Gestational Diabetes Mellitus." *Diabetes Care* 32, no. 12 (2009): 2236–41.
5. Lenny R. Vartanian, Marlene B. Schwartz, and Kelly D. Brownell. "Effects of Soft Drink Consumption on Nutrition and Health: A Systematic Review and Meta-Analysis." *American Journal of Public Health* 97 (2007): 667–75.
6. William Reymond. *Toxic: Obésité, malbouffe, maladie: enquête sur les vrais coupables*. Paris: Flammarion, 2007. p. 338.
7. Centers for Disease Control and Prevention, "National Diabetes Fact Sheet," 2007. http://www.cdc.gov/diabetes/pubs/pdf/ndfs_2007.pdf.
8. Claudia Sanmartin and Jason Gilmore. "Diabetes: Prevalence and Care." *Statistics Canada Catalogue no. 82-003-X* (Ottawa: Statistics Canada, 2008). http://www.statcan.gc.ca/pub/82-003-x/2008003/article/10663-eng.pdf.

9. François Reeves. *Prévenir l'infarctus ou y survivre*. Montreal: Éditions MultiMondes and Ste-Justine, 2007.
10. G.A. Bray, S.J. Nielsen, and B.M. Popkin. "Consumption of High-Fructose Corn Syrup in Beverages May Play a Role in the Epidemic of Obesity." *American Journal of Clinical Nutrition* 79 (2004): 537–43.
11. Ravi Dhingra et al. "Soft Drink Consumption and Risk of Developing Cardiometabolic Risk Factors and the Metabolic Syndrome in Middle-aged Adults in the Community." *Circulation* 116 (2007): 480–88.
12. J. Montonen et al. "Consumption of Sweetened Beverages and Intakes of Fructose and Glucose Predict Type 2 Diabetes Occurrence." *Journal of Nutrition* 137 (2007): 1447–92.
13. L. Chen et al. "Reducing Consumption of Sugar-sweetened Beverages Is Associated with Reduced Blood Pressure. A Prospective Study among United States Adults." *Circulation* 121 (2010): 2398–2406.
14. T.T. Fung et al. "Sweetened Beverage Consumption and Risk of Coronary Heart Disease in Women." *American Journal of Clinical Nutrition* 89 (2009): 1037–42.
15. K.L. Tucker et al. "Colas, but Not Other Carbonated Beverages, Are Associated with Low Bone Mineral Density in Older Women: The Framingham Osteoporosis Study." *American Journal of Clinical Nutrition* 84 (2006): 936–42.
16. L. Libuda et al. "Association between Long-Term Consumption of Soft Drinks and Variables of Bone Modeling and Remodeling in a Sample of Healthy German Children and Adolescents." *American Journal of Clinical Nutrition* 88 (2008): 1670–77.
17. K.A. Page et al. "Effects of Fructose vs. Glucose on Regional Cerebral Blood Flow in Brain Regions Involved with Appetite and Reward Pathways." *Journal of the American Medical Association* 309, no. 1 (2013): 63–70.
18. Guy Fagherazzi et al. "Consumption of Artificially and Sugar Sweetened Beverages and Incident Type 2 Diabetes in the E3N-EPIC Cohort." *American Journal of Clinical Nutrition*. Available at doi:10.3945/ajcn.112.050997.
19. J. Kiyah Duffey et al. "Food Price and Diet and Health Outcomes—20 Years of the CARDIA Study." *Archives of Internal Medicine* 170, no. 5 (2010): 420–26.
20. Coca-Cola China. *Live Positively: Coca-Cola China 2008/09 Sustainability Review*. Beijing: Coca-Cola Company, 2009. Available at http://

assets.coca-colacompany.com/16/60/e69bd11e47f4978b02cf9753f170/EN_final_draft_for_web.pdf.

21. Cléa Desjardins. "Weight Loss Today Keeps the Doctor Away." Concordia University. http://www.concordia.ca/now/media-relations/news-releases/20120716/weight-loss-today-keeps-the-doctor-away.php. Posted July 16, 2012.

CHAPTER SIX

1. "Fondation des maladies du coeur: Les ministres de la Santé doivent prendre des décisions quant à la réduction du sodium." fmcoeur.ca. http://www.fmcoeur.com/site/apps/nlnet/content2.aspx?c=ntJXJ8MMIqE&b=3562731&ct=11520003. Posted November 26, 2011.
2. Kirsten Bibbins-Domingo et al. "Projected Effects of Dietary Salt Reductions on Future Cardiovascular Disease." *New England Journal of Medicine* 362, no. 7 (2010): 590–97.
3. W. Willett and A. Ascherio. "Trans Fatty Acids. Are the Effects only Marginal?" *American Journal of Public Health* 84, no. 5 (1994): 722–24.
4. Ibid.
5. Trans Fat Task Force. "TRANSforming the Food Supply: Report of the Trans Fat Task Force." Ottawa: Health Canada, 2006. Available at http://hc-sc.gc.ca/fn-an/nutrition/gras-trans-fats/tf-ge/tf-gt_rep-rap-eng.php.
6. "Canada's New Government Calls on Industry to Adopt Limits for Trans Fat." Health Canada. Press release, June 20, 2007. Available at http://www.docstoc.com/docs/3490814/Health-Canada-News-Release---Trans-Fats-June-20-2007-httpwww-.
7. "Eliminating Trans Fat." Heart and Stroke Foundation. http://www.heartandstroke.qc.ca/site/c.pkI0L7MMJrE/b.3660121/k.51B9/Eliminating_trans_fat.htm. Posted January 2010.
8. "FDA takes steps to further reduce trans fats in processed foods." U.S. Food and Drug Administration. Press release, November 7, 2013. Available at http://www.fda.gov/NewsEvents/Newsroom/PressAnnouncements/ucm373939.htm.
9. Melissa Walton-Shirley. "The McDonald's Frappé: Warn your patients of this latest example of food-industry terrorism." theheart.org. http://blogs.theheart.org/melissa-walton-shirley-blog/2010/5/10/the-mcdonald-s-frappe-warn-your-patients-of-this-latest-example-of-food-industry-terrorism. Posted May 10, 2010.

10. Ibid.
11. Emily Ferenczi et al. "Can a Statin Neutralize the Cardiovascular Risk of Unhealthy Dietary Choices?" *American Journal of Cardiology* 106 (2010): 587–92.
12. Ibid.
13. É. Counil et al. *Trans-Polar Fat 2008: An Update on Atherogenic Affects and Regulatory issues in Nunavik*. December 9, 2008. ArcticNet Student Day, Quebec.
14. É. Counil et al. "Sugar-sweetened Beverages and the Metabolic Syndrome in the Inuit of Northern Quebec." December 11, 2008. 2008 Arctic Change Conference, Quebec.
15. É. Counil et al. *Trans-Polar Fat 2008: An Update on Atherogenic Effects and Regulatory Issues in Nunavik*. December 9, 2008. ArcticNet Student Day, Quebec.
16. É. Counil et al. "Consumption of Sugar-sweetened Beverages and Components of the Metabolic Syndrome in Inuit Adults of Northern Québec." 2008 Arctic Change Conference, December 11, 2008.
17. Jean Hamann. "La fin des gras trans au Nunavik?" *Le Fil*. http://www.aufil.ulaval.ca/articles/fin-des-gras-trans-nunavik-7512.html. Posted April 17, 2008.

CHAPTER SEVEN

1. "Thousands of Barbie Accessory Toys Recalled after Lead Violation." CBC News, http://www.cbc.ca/news/world/story/2007/09/04/mattel-recall.html. Updated September 5, 2007.
2. Adam K. Rowden et al. "Lead Toxicity." Medscape Reference. Available at http://emedicine.medscape.com/article/1174752-overview.
3. Rick Nevin. "How Lead Exposure Relates to Temporal Changes in IQ, Violent Crime, and Unwed Pregnancy." *Environmental Research* 83, no. 1 (2000): 1–22.
4. Huo Wenmian, Yao Tandong, and Li Yuefang. "Increasing Atmospheric Pollution Revealed by Pb Record of a 7000-m Ice Core." *Chinese Science Bulletin* 44 (1999): 14.

CHAPTER EIGHT

1. Al Gore. *Earth in the Balance: Ecology and the Human Spirit*. New York: Houghton Mifflin, 1992.

2. Al Gore. *An Inconvenient Truth: The Planetary Emergency of Global Warming and What We Can Do about It.* Emmaus, Pennsylvania: Rodale Press, 2006.
3. National Oceanic and Atmospheric Administration. "NOAA: Carbon dioxide levels reach milestone at Arctic sites." NOAA. http://researchmatters.noaa.gov/news/Pages/arcticCO2.aspx. Posted May 31, 2012.
4. Benjamin Santer et al. *Global Climate Change Impacts in the United States.* Washington: U.S. Global Change Research Program, 2009. Available at http://nca2009.globalchange.gov.
5. "A la conquête du feu." Homonidés.com. http://www.hominides.com/html/exposition/conquete-du-feu-terra-amata-nice-0092.php. Accessed May 3, 2013.
6. A quirk of medicine: The legend of Prometheus provides a hint that the Greeks suspected the liver to be one of the few human organs to spontaneously regenerate when injured. A valuable and surprising premonition on their part, as this characteristic of the liver is what makes live liver transplants possible. The section of liver that is removed from the donor will regrow after the transplant. Just as siblings can donate a kidney to each other, they can donate part of the liver to one another knowing that it will regenerate in the donor. Researchers are hoping that stem cells will one day make it possible to donate parts of other organs such as the heart.
7. Niles Eldredge. "The Sixth Extinction." ActionBioscience.org. http://www.actionbioscience.org/newfrontiers/eldredge2.html. Posted June 2001.
8. Albert Jacquard. *La Légende de la vie.* Paris: Flammarion, 1999.
9. John C. Avise, Stephen P. Hubbell, and Francisco J. Ayala. "In the Light of Evolution II: Biodiversity and Extinction." *Proceedings of the National Academy of Sciences* 105, no. 1 (2005): 11453–57. Available at www.pnas.org/content/105/suppl.1/11453.full.
10. Paul R. Ehrlich and Robert M. Pringle. "Where Does Biodiversity Go from Here? A Grim Business-as-usual Forecast and a Hopeful Portfolio of Partial Solutions." *Proceedings of the National Academy of Sciences* 105, no. Supplement 1 (2008): 11579–86. Available at http://www.pnas.org/content/105/suppl.1/11579.full.
11. Ibid.
12. "A la conquête du feu." Homonidés.com. http://www.hominides.com/

html/exposition/conquete-du-feu-terra-amata-nice-0092.php. Accessed May 3, 2013.

13. "Situation mondiale de l'énergie." L'Association Française de l'Hydrogène. http://www.afh2.org/uploads/memento/Fiche2.1revisee.pdf. Updated February 20, 2007.
14. David Suzuki with Faisal Moola. "Putting Humans in Their Place." *Science Matters*. http://www.davidsuzuki.org/blogs/science-matters/2008/10/putting-humans-in-their-place/.

CHAPTER NINE

1. Salim Yusuf et al. "Effect of Potentially Modifiable Risk Factors Associated With Myocardial Infarction in 52 Countries (the InterHeart Study): Case-Control Study." *Lancet* 364 (2004): 937–52.
2. Anne Milan. "Mortality: Causes of Death, 2007." Statistics Canada (2007), http://www.statcan.gc.ca/pub/91-209-x/2011001/article/11525-eng.htm. Updated July 20, 2011.
3. Salim Yusuf. "Two Decades of Progress in Preventing Vascular Disease." *Lancet* 360, no. 9326 (2002): 2–3.
4. Claude Lenfant, "Review of NIH Programs with an Emphasis on Stroke, Heart Disease, and Blood Disease." Statement before the House Committee on Energy and Commerce Subcommittee on Health, June 6, 2002. Available at http://www.hhs.gov/asl/testify/t020606.html.
5. World Health Organization. *Atlas of Health in Europe*, 2nd edition. Geneva: World Health Organization, 2008.
6. John D. Cantwell. "Cardiovascular Disease and Olympic Games in China." *American Journal of Cardiology* 101, no. 4 (208): 542–43.
7. Ibid.
8. Ross McKitrick. "Why Did U.S. Air Pollution Decline after 1970?" *Empirical Economics* 33 (2007): 491–513.
9. R.D. Fraser. "Average Annual Number of Deaths and Death Rates for Leading Causes of Death, Canada, For Five-Year Periods, 1921 to 1974. Series B35–50." *Section B: Vital Statistics and Health*. Statistics Canada. http://www.statcan.gc.ca/pub/11-516-x/sectionb/4147437-eng.htm. Updated October 22, 2008.
10. World Bank. Cost of Pollution in China: Economic Estimates of Physical Damages (2007). http://web.worldbank.org/WBSITE/EXTERNAL/COUNTRIES/EASTASIAPACIFICEXT/EXTEAPREGTOPENVIRONMENT/

0,,contentMDK:21252897~pagePK:34004173~piPK:34003707~theSitePK:502886,00.html.

11. Jacqueline Muller-Nordhorn et al. "An Update on Regional Variation in Cardiovascular Mortality within Europe." *European Heart Journal*. Available at doi:10.1093/eurheartj/ehm604.
12. "Smog." *Absolute Astronomy*, www.absoluteastronomy.com/topics/Smog#encyclopedia. Accessed May 4, 2013.
13. Greater London Authority. *50 Years On: The Struggle for Air Quality in London since the Great Smog of December 1952*. London: Greater London Authority, 2002. Available at http://legacy.london.gov.uk/mayor/environment/air_quality/docs/50_years_on.pdf.
14. Michelle L. Bell, Devra L. Davis, and Tony Fletcher. "A Retrospective Assessment of Mortality from the London Smog Episode of 1952: The Role of Influenza and Pollution." *Environmental Health Perspectives* 112, no. 1 (2004): 6–8.
15. "Air Pollution: Changing Air Quality & Clean Air Acts." Envirupedia. http://www.air-quality.org.uk/03.php.
16. K.C. Heidorn. "A Chronology of Important Events in the History of Air Pollution Meteorology to 1970." *Bulletin of the American Meteorology Society* 59, no. 12 (1978): 1589.
17. Jean-Marie Robine. "Projet Canicule." INSERM. http://ec.europa.eu/health/ph_projects/2005/action1/action1_2005_full_en.htm.
18. Jean-Marie Robine et al. *Étude de l'impact de la canicule d'août 2003 sur la population européenne [The 2003 Heat Wave Project]*. Montpelier: Canicule Project, 2007. Available at http://ec.europa.eu/health/ph_projects/2005/action1/docs/action1_2005_inter_15_en.pdf.
19. Martine Bungener. "Canicule estivale: la triple vulnérabilité des personnes âgées [Summer heat wave: the triple vulnerability of the elderly]." *Mouvements* 32 (2004).
20. Denis Hémon et al. "Excess Mortality Related to the August 2003 Heat Wave in France." *International Archives of Occupational and Environmental Health* 80 (2006): 16–24.
21. U.S. Global Change Research Program. *Global Climate Change Impacts in the United States: Second National Climate Change Assessment*. New York: Cambridge University Press, 2009. Available at http://downloads.globalchange.gov/usimpacts/pdfs/climate-impacts-report.pdf.

CHAPTER TEN

1. Robert D. Brook et al. "Particulate Matter Air Pollution and Cardiovascular Disease. An Update to the Scientific Statement from the American Heart Association." *Circulation* 121 (2010): 2331–78.
2. D.W. Dockery et al. "An Association between Air Pollution and Mortality in Six U.S. Cities." *New England Journal of Medicine* 329 (1993): 1753–59. Available at http://www.nejm.org/doi/full/10.1056/NEJM199312093292401.
3. Elaine Appleton Grant. "Prevailing Winds." *HSPH News*, Fall 2012. Available at http://www.hsph.harvard.edu/news/magazine/f12-six-cities-environmental-health-air-pollution/.
4. Annette Peters. "Air Quality and Cardiovascular Health: Smoke and Pollution Matter." *Circulation* 120 (2009): 924–27. Available at http://circ.ahajournals.org/content/120/11/924.full.
5. Qinghua Sun et al. "Long-Term Air Pollution Exposure and Acceleration of Atherosclerosis and Vascular Inflammation in an Animal Model." *Journal of the American Medical Association* 294 (2005): 3003–10. Available at jama.jamanetwork.com/article.aspx?articleid=202049.
6. Aruni Bhatnagar. "Environmental Cardiology: Studying Mechanistic Links between Pollution and Heart Disease." *Circulation Research* 99 (2006): 692–705.
7. A. Le Tertre et al. "Short-Term Effects of Particulate Air Pollution on Cardiovascular Disease in Eight European Cities." *Journal of Epidemiology and Community Health* 56 (2002): 773–9. Available at http://jech.bmj.com/content/56/10/773.full.
8. B. Hoffmann et al. "Residential Exposure to Traffic Is Associated with Coronary Atherosclerosis." *Circulation* 116 (2007): 489–96. Available at http://circ.ahajournals.org/content/116/5/489.full.
9. Andrea Baccarelli et al. "Living near Major Traffic Roads and Risk of Deep Vein Thrombosis." *Circulation* 119 (2009): 3118–24. Available at http://circ.ahajournals.org/content/119/24/3118.full.
10. Annette Peters et al. "Increased Particulate Air Pollution and the Triggering of Myocardial Infarction." *Circulation* 103 (2001): 2810–15. Available at http://circ.ahajournals.org/content/103/23/2810.ful.
11. Annette Peters et al. "Exposure to Traffic and the Onset of Myocardial Infarction." *New England Journal of Medicine* 351 (2004):

1721–30. Available at http://www.nejm.org/doi/full/10.1056/NEJMoa040203#t=article

12. Gerard Hoek et al. "Association between Mortality and Indicators of Traffic-related Air Pollution in the Netherlands: A Cohort Study." *Lancet* 360 (2002): 1203–09. Abstract available at http://www.thelancet.com/journals/lancet/article/PIIS0140673602112803/abstract.
13. Jaana Kettunen et al. "Associations of Fine and Ultrafine Particulate Air Pollution with Stroke Mortality in an Area of Low Air Pollution Levels." *Stroke* 38 (2007): 918–22. Available at http://stroke.ahajournals.org/content/38/3/918.long
14. Gregory A. Wellenius et al. "Ambient Air Pollution and the Risk of Acute Ischemic Stroke." *Archives of Internal Medicine* 172, no. 3 (2012): 229–34. Available at http://archinte.jamanetwork.com/article.aspx?articleid=1108717.
15. Douglas W. Dockery et al. "Association of Air with Increased Incidence of Ventricular Tachyarrhythmias Recorded by Implanted Cardioverter Defibrillators." *Environmental Health Perspectives* 113, no. 6 (2005): 670–74. Available at http://www.ncbi.nlm.nih.gov/pmc/articles/PMC1257589/.
16. François Reeves. *Prévenir l'infarctus ou y survivre.* Montreal: Multimondes/Éditions du CHU Sainte-Justine, 2007. p. 82.
17. C. Arden Pope III et al. "Lung Cancer, Cardiopulmonary Mortality, and Long-term Exposure to Fine Particulate Air Pollution." *Journal of the American Medical Association* 287, no. 9 (2002): 1132–41. Available at http://jama.jamanetwork.com/article.aspx?articleid=194704.
18. C. Arden Pope III, M. Ezzati, and D.W. Dockery. "Fine-Particulate Air Pollution and Life Expectancy in the United States." *New England Journal of Medicine* 360 (2009): 376–86. Available at http://www.nejm.org/doi/full/10.1056/NEJMsa0805646#t=article.
19. C. Arden Pope III et al. "Cardiovascular Mortality and Long-Term Exposure to Particulate Air Pollution: Epidemiological Evidence of General Pathophysiological Pathways of Disease." *Circulation* 109 (2004): 71–77. Available at http://circ.ahajournals.org/content/109/1/71.long.
20. Michelle L. Bell et al. "Emergency Hospital Admissions for Cardiovascular Diseases and Ambient Levels of Carbon Monoxide: Results for 126 United States Urban Counties, 1999–2005." *Circulation* 120 (2009): 949–55. Available at http://circ.ahajournals.org/content/120/11/949.full.
21. Committee on Estimating Mortality Risk Reduction Benefits from

Decreasing Tropospheric Ozone Exposure, National Research Council. *Estimating Mortality Risk Reduction and Economic Benefits from Controlling Ozone Air Pollution*. Washington: The National Academies Press, 2008.

22. Frank O'Donnell. "Earth Day Notes." Clean Air Watch. http://www.cleanairwatch.org/2008/04/earth-day-notes-nas-rebukes-bush-on.html. Posted April 22, 2008.
23. Robert D. Brook, et al. "Particulate Matter Air Pollution and Cardiovascular Disease. An Update to the Scientific Statement From the American Heart Association." *Circulation* 121 (2010): 2331–78. Available at http://circ.ahajournals.org/content/121/21/2331.full.
24. World Health Organization. "World Health Report, 2002." World Health Organization, http://www.who.int/whr/2002/en/.
25. C. Arden Pope III. "The Expanding Role of Air Pollution in Cardiovascular Disease." *Circulation* 119 (2009): 3050–52. Available at http://circ.ahajournals.org/content/119/24/3050.full#ref-11.
26. World Health Organization. "How Reducing Short-Lived Climate Pollutants Can Lower the Death Toll." World Health Organization, http://www.who.int/hia/green_economy/reducing_air_pollution/en/.html.
27. A. Prüss-Üstün and C. Corvalán. *Preventing Disease through Healthy Environments. Towards an Estimate of the Environmental Burden of Disease*. Geneva: World Health Organization, 2006.
28. T.J. Grahame and R.B. Schlesinger. "Cardiovascular Health and Particulate Vehicular Emissions: A Critical Evaluation of the Evidence." *Air Quality Atmospheric Health* 3 (2010): 23–27. Available at http://link.springer.com/article/10.1007%2Fs11869-009-0047-x#page-1.
29. David R. Boyd and Stephen J. Genuis. "The Environmental Burden of Disease in Canada: Respiratory Disease, Cardiovascular Disease, Cancer, and Congenital Affliction." *Environmental Research* 106 (2008): 240–49. Abstract available at http://www.sciencedirect.com/science/article/pii/S0013935107001600.
30. B. Urch et al. "Relative Contributions of $PM_{2.5}$ Chemical Constituents to Acute Arterial Vasoconstriction in Humans." *Inhalation Toxicology* 16 (2004): 345–52. Abstract available at http://informahealthcare.com/doi/abs/10.1080/08958370490439489.
31. B. Urch et al. "Acute Blood Pressure Responses in Healthy Adults during Controlled Air Pollution Exposures." *Environmental Health Perspectives*

113, no. 8 (2005): 1052–55. Available at http://www.ncbi.nlm.nih.gov/pmc/articles/PMC1280348/.

32. C.R. Bartoli et al. "Mechanisms of Inhaled Fine Particulate-induced Arterial Blood Pressure Changes." *Environmental Health Perspectives* 117 (2009): 361–66. Available at http://www.ncbi.nlm.nih.gov/pmc/articles/PMC2661904/.
33. B. Jain Nitin et al. "Lead Levels and Ischemic Heart Disease in a Prospective Study of Middle-aged and Elderly Men: The VA Normative Aging Study." *Environmental Health Perspectives* 115, no. 6 (2006): 871–75. Available at http://www.ncbi.nlm.nih.gov/pmc/articles/PMC1892138/.
34. Robert D. Brook. "You Are What You Breathe: Evidence Linking Air Pollution and Blood Pressure." *Current Hypertension Reports* 7 (2005): 427–34. Abstract available at http://www.unboundmedicine.com/medline/citation/16386198/You_are_what_you_breathe:_evidence_linking_air_pollution_and_blood_pressure.
35. Alan H. Lockwood. "Diabetes and Air Pollution." *Diabetes Care* 25 (2002): 1487–88. Available at http://care.diabetesjournals.org/content/25/8/1487.full.
36. Andrew Rundle et al. "Association of Childhood Obesity With Maternal Exposure to Ambient Air Polycyclic Aromatic Hydrocarbons During Pregnancy." *American Journal of Epidemiology* 175, no. 11 (2012): 1163–72. Abstract available at http://aje.oxfordjournals.org/content/175/11/1163.
37. A.E. Silverstone et al. "Polychlorinated Biphenyl (PCB) Exposure and Diabetes: Results from the Anniston Community Health Survey." *Environmental Health Perspectives* 120, no. 5 (2012): 727–32. Available at http://www.ncbi.nlm.nih.gov/pmc/articles/PMC3346783/.
38. "Crop Fungicide Linked to Diabetes." UPI.com. http://www.upi.com/Health_News/2011/06/26/Crop-fungicide-linked-to-diabetes/UPI-87841340728426/. Posted June 26, 2011.
39. T.F. Bateson and J. Schwartz. "Who Is Sensitive to the Effects of Particles on Mortality? A Case-Crossover Analysis." *Epidemiology* 15 (2004): 143–49. Available at http://www.apexepi.com/download/Epidemiology%20 2004.pdf.
40. A. Zanobetti and J. Schwartz. "Are Diabetics More Susceptible to the Health Effects of Airborne Particles?" *American Journal of Respiratory Critical Care Medicine* 164, no. 5 (2001): 831–33. Available at http://

www.atsjournals.org/doi/full/10.1164/ajrccm.164.5.2012039?journalCode=ajrccm&.

41. A. Zanobetti and J. Schwartz. "Cardiovascular Damage by Airborne Particles: Are Diabetics More Susceptible?" *Journal of Epidemiology* 13 (2002): 588–92. Available at www.hsph.harvard.edu/clarc/pubs/endnote103-zanobetti.pdf.
42. Robert D. Brook. "You Are What You Breathe: Evidence Linking Air Pollution and Blood Pressure." *Current Hypertension Reports* 7 (2005): 427–34. Abstract available at http://www.unboundmedicine.com/medline/citation/16386198/You_are_what_you_breathe:_evidence_linking_air_pollution_and_blood_pressure.
43. Jennifer Weuve et al. "Exposure to Particulate Air Pollution and Cognitive Decline in Older Women." *Archives of Internal Medicine* 172, no. 3 (2012): 219–27. Available at http://archinte.jamanetwork.com/article.aspx?articleid=1108716.
44. E. van Wijngaarden, J.R. Campbell, and D.A. Cory-Slechta. "Bone Lead Levels Are Associated with Measures of Memory Impairment in Older Adults." *Neurotoxicology* 30, no. 4 (2009): 572–80. Available at http://www.ncbi.nlm.nih.gov/pmc/articles/PMC2719051/.
45. David R. Boyd and Stephen J. Genuis. "The Environmental Burden of Disease in Canada: Respiratory Disease, Cardiovascular Disease, Cancer, and Congenital Affliction." *Environmental Research* 106 (2008): 240–49. Abstract available at http://www.sciencedirect.com/science/article/pii/S0013935107001600.
46. Ibid.
47. World Bank and The Government of the People's Republic of China. *Cost of Pollution in China: Economic Estimates of Physical Damages.* Beijing: State Environmental Protection Agency, Peoples' Republic of China, 2007. Available at http://siteresources.worldbank.org/INTEAPREGTOPENVIRONMENT/Resources/China_Cost_of_Pollution.pdf.
48. "Tackling the Global Clean Air Challenge." WHO press release, September 26, 2011. Available at http://www.who.int/mediacentre/news/releases/2011/air_pollution_20110926/en/index.html.
49. United Nations. "Dangers of Air Pollution Worse than Previously Thought, UN Health Agency Warns." UN News Centre. http://www.un.org/apps/news/story.asp?NewsID=44586&Cr=pollution&Cr1=#.UYbtLStASbw. Posted April 8, 2013.

50. OECD. *OECD Environmental Outlook to 2050*. Paris: OECD Publishing, 2012. Available at http://dx.doi.org/10.1787/9789264122246-en.

CHAPTER ELEVEN

1. Salim Yusuf et al. "Effect of Potentially Modifiable Risk Factors Associated with Myocardial Infarction in 52 Countries (the InterHeart Study): Case-Control Study." *Lancet* 364 (2004): 937–52.
2. Committee on Secondhand Smoke Exposure and Acute Coronary Events. *Secondhand Smoke Exposure and Cardiovascular Effects: Making Sense of the evidence*. Washington: Institute of Medicine. National Academy of Science, 2010.
3. Michael S. Friedman et al. "Impact of Changes in Transportation and Commuting Behaviors during the 1996 Summer Olympic Games in Atlanta on Air Quality and Childhood Asthma." *Journal of the American Medical Association* 285, no. 7 (2001): 897–905. Available at http://jama.jamanetwork.com/article.aspx?articleid=193572.
4. John D. Cantwell. "Cardiovascular Disease and Olympic Games in China." *America Journal of Cardiology* 101, no. 4 (2008): 542–43.
5. Xing Wang et al. "Evaluating the Air Quality Impacts of the 2008 Beijing Olympic Games: On-Road Emission Factors and Black Carbon Profiles." *Atmospheric Environment* 43 (2009): 4535–43.
6. L. Clancy et al. "Effect of Air-Pollution Control on Death Rates in Dublin, Ireland: An Intervention Study." *Lancet* 360, no. 9341 (2002): 1210–14.
7. C. Arden Pope III, Majid Ezzati, and Douglas W. Dockery. "Fine-Particulate Air Pollution and Life Expectancy in the United States." *New England Journal of Medicine* 360 (2009): 376–86.
8. James McCreanor et al. "Respiratory Effects of Exposure to Diesel Traffic in Persons with Asthma." *New England Journal of Medicine* 357 (2007): 2348–58.
9. United Nations Environment Programme. "Wide Spread and Complex Climatic Changes Outlined in New UNEP Project Atmospheric Brown Cloud Report—Cities Across Asia Get Dimmer: Impacts on Glaciers, Agriculture and the Monsoon Get Clearer." UNEP. http://www.unep.org/documents.multilingual/default.asp?documentid=550&articleid=5978&l=en. Posted November 13, 2008.

CHAPTER TWELVE

1. Y. Beaudouin and F. Cavayas. *Projet Biotopes. Évolution des occupations du sol, du couvert végétal et des îlots de chaleur sur le territoire*

de la Communauté métropolitaine de Montréal (1984–2005). Montreal: Département de géographie de l'Université de Montréal et Département de géographie de l'Université du Québec à Montréal, January 2008.
2. U.S. Global Change Research Program. *Global Climate Change Impacts in the United States*. New York: Cambridge University Press, 2009. Available at http://www.globalchange.gov/publications/reports/scientific-assessments/us-impacts/global-climate-change.
3. D. Boulet and S. Melançon. *Environmental Assessment Report. Air Quality in Montreal. Montreal: Direction de l'environnement et du développement durable, 2010*. Available at http://ville.montreal.qc.ca/pls/portal/docs/page/enviro_fr/media/documents/rsqa_report_2010_en.pdf.
4. Gregory A. Wellenius et al. "Ambient Air Pollution and the Risk of Acute Ischemic Stroke." *Archives of Internal Medicine* 172, no. 3 (2012): 229–34.
5. Mathieu-Robert Sauvé. "Montréal se réchauffe dangereusement!" *Forum*, http://nouvelles.umontreal.ca/archives/2007-2008/content/view/1112/327/index.html. Posted March 10, 2008.
6. E.M. Fischer and C. Schär. "Consistent Geographical Patterns of Changes in High-Impact European Heatwaves." *Nature Geoscience* 3 (2010): 398–403.

CHAPTER THIRTEEN

1. Catrine Tudor-Locke and David R. Bassett Jr. "How Many Steps per Day Are Enough? Preliminary Pedometer Indices for Public Health." *Sports Medicine* 34, no. 1 (2004): 1–8.
2. Peter Nielsen et al. "Primary Angioplasty versus Fibrinolysis in Acute Myocardial Infarction: Long-Term Follow-Up in the Danish Acute Myocardial Infarction 2 Trial." *Circulation* 121, no. 13 (2010): 1484–91. Available at doi:10.1161/CIRCULATIONAHA.109.873224.
3. Gehl Architects. http://www.gehlarchitects.com. Accessed May 6, 2013.
4. Jan Gehl. *Cities for People*. Washington: Island Press, 2010.
5. "Towards a green strategy." GardenVisit. http://www.gardenvisit.com/landscape_architecture/london_landscape_architecture/landscape_planning_pos_public_open_space/towards_a_green_strategy_for_london_turner. Posted 2008.
6. Georges-Eugène (Baron) Haussmann is celebrated for having designed many of the great monuments, boulevards, parks, and public buildings of late-nineteenth-century Paris and for bringing order and vision to a

hitherto haphazard and unmanageable city layout. Although executing Haussmann's plan necessitated demolishing many existing structures, the result was a more stately and harmonious city center. Baron Haussmann inspired a whole new generation of urban planners whose work may be seen in numerous cities, including Washington, D.C.

7. TechnoParc Montréal. http://www.technoparc.com. Accessed May 6, 2013.
8. "Site Outremont." Université de Montréal. http://www.umontreal.ca/grandsprojets/outremont/projet/index.html. Accessed May 6, 2013.
9. Fiducie de recherche sur la forêt des Cantons-de-l'Est / Eastern Townships Forest Research Trust. http://www.frfce.qc.ca/FFiduciaires.html. Accessed May 6, 2013.

CHAPTER FIFTEEN

1. L. Jacques Ménard, letter to the editor, *La Presse*, February 12, 2009.
2. International Union for Conservation of Nature. *2007/2008 IUCN Red List of Threatened Species*. International Union for Conservation of Nature, SSC Red List Programme, Cambridge. Available at: http://www.iucnredlist.org.
3. Jean Giono. *The Man Who Planted Trees*. Chelsea, Vermont: Chelsea Green Publishing, 1985.
4. United Nations Environment Programme. "The Economics of Ecosystems and Biodiversity." *UNEP Annual Report: Seizing the Green Opportunity*. Nairobi: UNEP, 2009. Available at http://www.unep.org/publications/ebooks/annual-report09/Content.aspx?id=ID0E2B1.
5. Sylvain Michelet. "Écopsychologie: la psy se met au vert." *Psychologies* May 2009.
6. Ann Sloan Devlin and Allison B. Arneill. "Health Care Environments and Patient Outcomes: A Review of the Literature." *Environment and Behavior* 35 (2003): 665.
7. Rachel Kaplan and Steven Kaplan. *The Experience of Nature: A Psychological Perspective*. Cambridge: Cambridge University Press, 1989.
8. R.S. Ulrich. "View through a Window May Influence Recovery from Surgery." *Science* 224 (1984): 420–21.
9. Jo Nurse et al. "An Ecological Approach to Promoting Population Mental Health and Well-Being: A Response to the Challenge of Climate Change." *Perspectives in Public Health* 130, no. 1 (2010): 27–33.
10. F.E. Kuo and W.C. Sullivan. "Aggression and Violence in the Inner City:

Effects of Environment via Mental Fatigue." *Environment and Behavior* 33 (2001): 543–71.

11. F.E. Kuo and W.C. Sullivan. "Environment and Crime in the Inner City: Does Vegetation Reduce Crime?" *Environment and Behavior* 33 (2001): 343–67.
12. F.E. Kuo et al. "Fertile Ground for Community: Inner-City Neighborhood Common Spaces." *American Journal of Community Psychology* 26 (1998): 823–51.
13. M.E. Austin. "Partnership Opportunities in Neighborhood Tree-planting Initiatives: Building from Local Knowledge." *Journal of Arboriculture* 28 (2002): 178–86.
14. L.M. Wilson. "Intensive Care Delirium: The Effect of Outside Deprivation in a Windowless Unit." *Archives of Internal Medicine* 130 (1972): 225–26.
15. William Bird. *Natural Thinking: A Report for the Royal Society for the Protection of Birds. Investigating the Links between the Natural Environment, Biodiversity, and Mental Health.* Bedfordshire: Royal Society for the Protection of Birds, 2007. Available at http://www.rspb.org.uk/ourwork/policy/health/index.asp.
16. Sylvie Van Brabant. *Earth Keepers: A Survival Guide for a Planet in Peril.* Montreal: National Film Board of Canada, 2009. http://www.rapideblanc.ca/en/films.html#visionnaires.
17. "The Billion Tree Campaign." United Nations Environment Programme. http://www.plant-for-the-planet-billiontreecampaign.org/. Accessed May 6 2013.
18. Jeffrey Monty. "Healthy Trees—Healthy Canadians." OHA Health Achieve 2004. Available at http://treecanada.ca/files/2813/4880/2942/OHA_Health_Acheive_2004.pdf.
19. David Suzuki and Wayne Grady. *Tree.* Vancouver: Greystone Books, 2004.
20. "The Billion Tree Campaign." United Nations Environment Programme. http://www.plant-for-the-planet-billiontreecampaign.org/. Accessed May 6 2013.
21. Quebec Adipose and Lifestyle Investigation in Youth (QUALITY). http://www.etudequalitystudy.ca/index.php?lang=en. Accessed May 6, 2013.
22. "Les enfants vivant à proximité d'espaces verts marchent davantage." *U de M Nouvelles.* http://www.nouvelles.umontreal.ca/recherche/sciences-de-la-sante/les-enfants-vivant-a-proximite-despaces-verts-marchent-davantage.html. Posted March 12, 2009.

23. http://treecanada.ca/en/resources/kids-teachers/tree trivia/.
24. T. Karl et al. "Efficient Atmospheric Cleansing of Oxidized Organic Trace Gases by Vegetation." *Science Express*. October 21, 2010. Available at http://www.sciencemag.org/content/suppl/2010/10/18/science.1192534.DC1/Karl-SOM.pdf.
25. L. O'Brien. "Strengthening Heart and Mind: Using Woodlands to Improve Mental and Physical Well-Being." *Unasylva* 57, no. 224 (2006): 56–61. Available at http://www.fao.org/docrep/009/a0789e/a0789e14.html.
26. Richard Mitchell and Frank Popham. "Effect of Exposure to Natural Environment on Health Inequalities: An Observational Population Study." *Lancet* 372 (2008): 1655–60.
27. Geoffrey H. Donovan et al. "The Relationship between Trees and Human Health. Evidence from the Spread of the Emerald Ash Borer." *American Journal of Preventive Medicine* 44, no. 2 (2013): 139–45.
28. Shirin-yoku Coalition. "What Is Shirin-yoku?" Shinrin-yoku: The Medicine of Being in the Forest. http://www.shinrin-yoku.org/shinrin-yoku.html. Accessed May 7, 2013.
29. Qing Li. "Effect of Forest Bathing Trips on Human Immune Function." *Environmental Health Prevention Medicine* 15 (2010): 9–17.
30. Bum Jin Park et al. "Physiological Effects of *Shinrin-yoku*: Evidence from Field Experiments in 24 Forests across Japan." *Environmental Health and Preventive Medicine* 15 (2010): 18–26.
31. Muriel Barbery. *The Elegance of the Hedgehog*. Translated by Alison Anderson. New York: Europa Editions, 2006. p. 169.

Image credits

Figure 1: Adapted from Al Gore. *Earth in the Balance: Ecology and the Human Spirit*. New York: Houghton Mifflin, 1992.

Figure 2: Adapted from C. Arden Pope III et al. "Cardiovascular Mortality and Exposure to Airborne Fine Particulate Matter and Cigarette Smoke: Shape of the Exposure-Response Relationship." *Circulation* 120 (2009): 941–48.

Figure 3: Adapted from R.D. Brook et al. "Air Pollution and Cardiovascular Disease: A Statement for Healthcare." *Circulation* 109 (2004): 2655–71. Available at http://sites.tufts.edu/cafeh/files/2012/01/air-pollution-and-cardiovascular-disease-a-statement.pdf.

Acknowledgments

I AM DEEPLY GRATEFUL to François Reeves, MD, for his vision and generosity of spirit; to Iva Cheung, Nancy Flight, and Rob Sanders of Greystone Books for their belief in and contributions to this book; to Tom Holzinger for his friendship and nonpareil editing skills; and to Kyle Irving-Moroz and Jim Higgins for being there.

JOAN IRVING

Index

The David Suzuki Foundation

THE DAVID SUZUKI FOUNDATION works through science and education to protect the diversity of nature and our quality of life, now and for the future.

With a goal of achieving sustainability within a generation, the Foundation collaborates with scientists, business and industry, academia, government and non-governmental organizations. We seek the best research to provide innovative solutions that will help build a clean, competitive economy that does not threaten the natural services that support all life.

The Foundation is a federally registered independent charity that is supported with the help of over 50,000 individual donors across Canada and around the world.

We invite you to become a member. For more information on how you can support our work, please contact us:

The David Suzuki Foundation
219–2211 West 4th Avenue
Vancouver, BC Canada V6K 4S2
www.davidsuzuki.org
contact@davidsuzuki.org
Tel: 604-732-4228 Fax: 604-732-0752

Checks can be made payable to the David Suzuki Foundation. All donations are tax-deductible.

Canadian charitable registration: (BN) 12775 6716 RR0001
U.S. charitable registration: #94-3204049